Is Hopping a Science?

Selected Topics of Hopping Conductivity

IS HOPPING A SCIENCE?

Selected Topics of Hopping Conductivity

Issai Shlimak

Bar-Ilan University, Israel

 **World Scientific**

NEW JERSEY · LONDON · SINGAPORE · BEIJING · SHANGHAI · HONG KONG · TAIPEI · CHENNAI

Published by

World Scientific Publishing Co. Pte. Ltd.
5 Toh Tuck Link, Singapore 596224
USA office: 27 Warren Street, Suite 401-402, Hackensack, NJ 07601
UK office: 57 Shelton Street, Covent Garden, London WC2H 9HE

British Library Cataloguing-in-Publication Data
A catalogue record for this book is available from the British Library.

IS HOPPING A SCIENCE?
Selected Topics of Hopping Conductivity

ISBN 978-981-4663-33-5

In-house Editor: Song Yu

Printed in Singapore

To my family

Preface

Hopping conductivity is a mechanism for charge transport for localized electrons. There is no contradiction between "conductivity" and "localization" for non-zero temperature. Hopping conductivity is an alternative to "normal" conductivity due to the motion of "free" non-localized carriers. The latter has been well known for a long time and is described in detail with different mechanisms of scattering and interaction in textbooks devoted to solid state physics, semiconductors and metals. The investigation of hopping conductivity does not have such a long history. The term "localization" for electrons in a lattice was first introduced in the early 1960s. The investigation of hopping conductivity has attracted much attention from both theorists and experimentalists. As a result, a biennial Conference was established, called "Hopping and Related Phenomena". The name has now been expanded to "Transport in Interacting Disordered Systems".

The title of this book derives from a curious incident that happened to me in 1989, when I arrived in the United States to participate in the "Hopping" conference, which was held that year in Chapel Hill, North Carolina. The immigration officer at JFK Airport in New York asked me the purpose of my visit. I replied: "Participation in an international conference". The officer then asked the topic of this conference and I answered: "Hopping and Related Phenomena". The officer was very surprised and said, "I never thought that hopping is a science".

Yes, hopping conductivity is an important science, consists of studying electron transport in important materials with a wide range of application in micro- and nano-electronics, including doped semiconductors and different disordered systems: glasses, amorphous films, granular materials, alloys, polymers, and even biological systems. Studying hopping conductivity is also closely connected with the investigation of many fundamental physical phenomena, such as electron–electron Coulomb interaction, spin interactions, many-body systems, percolation, noise, mesoscopics, small polarons, quantum oscillations and many others.

It was first shown by P. W. Anderson that the localization of electrons in a crystalline lattice may occur because of disorder. Electrons can be localized on impurities, vacancies, and other lattice imperfections, in the minima of random potential fluctuations arising from the random distribution of charged impurities in the solid, due to the heterogeneity of surfaces and interfaces. There are inherently disordered solids, such as amorphous semiconductors, glasses, granular metals, alloys, polymers, etc. In practice, however, even crystalline solids are somewhat disordered because of the random potential caused by chaotically distributed charged impurities and other natural defects. If the electron energy is much larger than the average amplitude of the random potential, electrons can move as "free" quasi-particles, as they do in metals. In non-degenerated semiconductors, however, electrons at low temperatures become localized in the potential wells, which leads to the hopping mechanism of transport at non-zero temperatures, as occurs in insulators. Thus, the problem of the metal–insulator transition is also connected with the investigation of hopping conductivity.

As an experimentalist, I have been involved in the study of hopping conductivity from the early stages. Most of my studies were carried out on a classical system — the impurity band in crystalline germanium doped with different shallow impurities. I observed how some problems relating to different aspects of hopping motion were studied and sometimes solved. This monograph is based on some of my own works devoted to better understanding of hopping conductivity. Most of these problems were at the center of research activity many years ago, and therefore, the papers referred to are not recent. This book is not a review or a textbook. The list of references may not be complete; only those papers are mentioned that are necessary to present and discuss the data.

The following problems will be considered:

1) Metal–insulator transition (MIT): critical indices of the MIT and quantitative analysis of delocalization process in the vicinity of the MIT.

2) Different types of hopping conductivity: nearest-neighbor hopping (NNH), variable-range hopping (VRH) without and with the existence of a soft Coulomb gap (CG) at the Fermi level; transformation of the hopping mechanism with a decrease of temperature; non-ohmic hopping conductivity.

3) "Hopping spectroscopy": specification of the density-of-states (DOS)

in the vicinity of the Fermi level, determination of the width of the CG; experimental discovering of the temperature induced smearing of the CG; detection of the mechanism of scatter of the impurity levels in solid-solution semiconductor compositions.

4) Nontrivial effects in hopping conduction: negative magnetoresistance, "magnetic hard gap" at the bottom of the DOS, electron–electron interaction-assisted hopping.

5) Hopping noise and electronic devices based on the hopping conductivity.

6) Interimpurity radiative recombination and hopping photoconductivity.

The choice of these topics is personal, and not exhaustive. The selected issues related to the problems in the study of which I personally participated and which were not described earlier in detail. Other issues are already covered in books and reviews, and therefore they are only mentioned briefly. In order that this selectivity would not hamper the perception of the entire problem, I tried to describe every topic from its beginning. I hope, this will make it clear to those who are not familiar with hopping science. In general, this monograph should be of interest to students and professionals in the field of semiconductor electronics.

I. Shlimak

Acknowledgments

I am grateful to colleagues for cooperation and long-term friendship, especially to Alexei Efros, Boris Shklovskii, Moshe Kaveh, Felix Gantmakher, Alexander Ionov, Mike Lee, Michael Pepper, Vladimir Ginodman, Greg Citver, Rolf Rentsch, Klaus-Jurgen Friedland, Sergey Baranovskii and Sergey Kravchenko. Special thanks go to Arkadii and Nataly Belostotsky who helped me overcome technical difficulties. I am also very thankful to Nathan Aviezer who read the manuscript and made useful improvements. And finally, I am thankful to my wife for her infinite patience.

Contents

Chapter 1

Critical Indices of the Metal–Insulator Transition

1.1 Introduction

The Metal–Insulator Transition (MIT) in doped semiconductors is a long-standing problem in condensed matter physics (see, for example, reviews [1, 2]. Characteristics of MIT in other systems are described in Ref. [3]. We start with some definitions.

For localized electrons, the wave function is concentrated in a limited region of space and decays exponentially with distance r:

$$\Psi(r) \propto \frac{1}{r} \exp\left(-\frac{r}{a_{\mathrm{B}}}\right). \tag{1.1}$$

The length a_{B} is called "radius of localization" or "Bohr radius" by analogy with the hydrogen atom whose wave function also decreases as $\exp(-r/a_{\mathrm{B}})$ with $a_{\mathrm{B}} = \hbar^2/(m_0 e^2)$, where e is the electron charge, m_0 is the electron mass, and the dielectric constant of the vacuum is assumed to be unity in units of the permittivity of free space $\kappa_{\mathrm{fs}} = 8.85 \times 10^{-12}\,\mathrm{F/m}$. In crystals, $\Psi(r)$ is modulated by the Bloch function having a lattice constant periodicity with a maximum at the lattice sites. Therefore, Eq. (1.1) describes the modulus of the envelope part of $\Psi(r)$ (Fig. 1.1(a)). In addition, for electrons localized on shallow impurities in semiconductors, the expression for the "radius of localization" includes the dielectric constant of the host material κ_0 and the effective mass of electrons m^*:

$$a_{\mathrm{B}} = \frac{\kappa_0 \hbar^2}{m^* e^2}. \tag{1.2}$$

In typical semiconductors, such as Si, Ge, $\kappa_0 \sim 10$ (in units of κ_{fs}) and $m^* \sim 0.1 m_0$. Therefore, in semiconductors, the value of a_{B} is two orders of magnitude larger than for the hydrogen atom and the wave function covers a volume containing thousands of crystal lattice nodes. In the vicinity of the

MIT, $\Psi(r)$ becomes more extended, but the exponential decay remains for a distance r outside the region of localization. This decay is characterized by the "localization length" ξ (Fig. 1.1(b)) which tends to infinity when approaching the critical point of the transition.

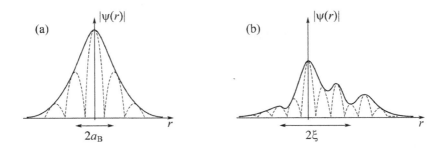

Fig. 1.1 One-dimensional schematic representation of the wave function for an electron localized on a single donor impurity (a) and in random-potential wells in the vicinity of the MIT (b).

In contrast, the wave function for delocalized electrons do not decay with increase of r. Just after the MIT, for barely metallic samples, $\Psi(r)$ is modulated, giving rise to a "correlation length" ξ which characterizes the average spatial scale of modulation (Fig. 1.2(a)). This is a similar parameter as the "localization length" on the insulating side of the MIT and also diverges to infinity when approaching the critical point of the MIT from the metallic side. Far from the MIT, deeply in metallic region, $\Psi(r)$ becomes the wave function typical of "free" electrons in a crystalline lattice with equal amplitude at all lattice sites (Fig. 1.2(b)).

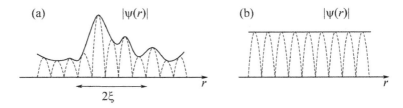

Fig. 1.2 One-dimensional schematic representation of the wave functions for a barely delocalized electron in the vicinity of the MIT (a) and for a "free" electron in a crystal (b).

At non-zero temperatures, almost all materials conduct electricity to some extent. The real difference between insulators and metals is revealed only at zero temperature. For metals, the electrical conductivity $\sigma(0)$ may be arbitrarily small, but non-zero, whereas for insulators, $\sigma(0)$ is exactly zero. This difference is due to the fact that the electron states at the highest occupied level (Fermi level E_F) in insulators are localized, whereas in metals, they are delocalized.

It was assumed for a long time that it is possible to distinguish between insulators and metals by means of the sign of $d\sigma/dT$ at low temperatures: for insulators, the electrical conductivity increases with temperature, $d\sigma/dT > 0$, whereas for metals, $d\sigma/dT < 0$. However, the study of quantum corrections to the conductivity [4] showed that under certain circumstances, $d\sigma/dT > 0$ even for metals. Therefore, this criterion is not valid.

For electron wave functions, the MIT corresponds to the transition from a delocalized to a localized state. Two main parameters influence the wave function in the ground state: electron–electron interactions and disorder. The MIT caused by interaction is called the Mott transition, whereas the transition induced by disorder in the system of non-interaction electrons, or when both factors are present simultaneously, is called the Anderson or Mott–Anderson transition.

There are many parameters that can lead to the MIT: impurities, magnetic field, pressure, electric field, etc. We denote them as x. Assume that the MIT occurs when one of these parameters increases. This means that at small x, below some critical value x_c, $\sigma(0) = 0$, whereas for $x > x_c$, $\sigma(0) \neq 0$. The question arises, what happens in the vicinity of the critical value x_c? Initially, the view of Mott dominated with a concept of "minimal metallic conductivity". This concept was based on the Drude expression for the conductivity of an electron gas of density n:

$$\sigma = \frac{e^2 k_F}{3\pi^2 \hbar} l, \qquad (1.3)$$

where l is the mean free path, and $k_F = (3\pi^2 n)^{1/3}$ is the wavenumber at the Fermi surface. The mean free path has an obvious minimum value (Ioffe–Regel limit): l cannot be smaller than the de Broglie wavelength $(1/k_F)$, which means that the three-dimensional (3d) conductivity cannot be smaller than the minimal value σ_M

$$\sigma_M(3d) = (3\pi^2)^{-2/3} \frac{e^2}{\hbar} n^{1/3} \approx k n^{1/3}, \qquad (1.4)$$

where $k = 0.25 \times 10^{-4} \, \text{Ohm}^{-1}$.

In two dimensions (2d), the expression for the "minimal metallic conductivity" contains only universal constants:

$$\sigma_M(2d) = C\frac{e^2}{\hbar}, \tag{1.5}$$

where C is a number independent of the specific model.

This concept implies a jump of $\sigma(0)$ at $x = x_c$ (Fig. 1.3, curve a). The concept of "minimal metallic conductivity" was tested experimentally many times. Despite the fact that the value of $\sigma(0)$ is determined by extrapolation and therefore the accuracy of such a determination is not very high, the experiments clearly show that the concept of σ_M is incorrect, and the transition is continuous.

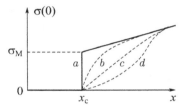

Fig. 1.3   Schematic representation of a MIT. a — Mott concept of "minimal metallic conductivity"; b–d — continuous transition $\sigma \propto (x - x_c)^\mu$ for $\mu < 1$ (b), $\mu = 1$ (c) and $\mu > 1$ (d).

The explanation of this discrepancy was given by the famous "scaling" hypothesis (see the short pioneering work [5] and review [6]). It has been shown that near the critical point of the transition, the value of the conductivity cannot be used for describing the electron transport. Electrical conductivity reveals itself in non-equilibrium conditions. Therefore, the description of the MIT as a phase transition requires a selection of functions describing electron transport in accordance with thermodynamics.

In scaling theory, the conductance G (Ohm^{-1}) of a cube of volume L^d is considered, where d is the number of space dimensions. One can introduce a dimensionless conductance $g = G(e^2/\hbar)^{-1}$, where $e^2/\hbar = 2.43 \times 10^{-4}$ Ohm^{-1} is called the "quantum of conductance". The relationship between conductance G, dimensionless conductance g and conductivity σ is given by

$$G = \sigma L^{d-2}, \qquad g = \left(\frac{e^2}{\hbar}\right)^{-1} \sigma L^{d-2}. \tag{1.6}$$

One of the main parameters of the phase transition is the correlation length ξ. This parameter is the characteristic size of the fluctuations assuming that $\xi \to \infty$ upon approaching the critical point of the transition x_c. Near the vicinity of the transition, different regions with insulating and conducting properties can exist simultaneously, and ξ is the average size of these regions. Therefore, the conductivity σ can be used as a basic parameter which characterizes the properties of a material only if the sample size L is larger than ξ. In the vicinity of the transition, ξ diverges, the condition $L > \xi$ is violated. Therefore, the connection between conductance G and conductivity σ will no longer be determined by the factor L^{d-2}. This is why G is chosen as the basic function to describe the MIT.

In the scaling model, a variable β is introduced, defined as the logarithmic derivative of the dimensionless conductance g with respect to the sample size L: $\beta = d\ln g/d\ln L = (L/g)(dg/dL)$. The scaling hypothesis postulates that the function $\beta(\ln g)$ is determined only by g and is a universal function for all space dimensions d. One can obtain the asymptotic parts of $\beta(\ln g)$ from the following consideration. At large g, the macroscopic relation (1.6) is valid, which gives $\beta = d-2$. At small g, when all the electrons are localized and their wave function (1.1) attenuates with localization radius a_B, the conductance of a sample of size L at zero temperature will be determined by the overlap of wave functions: $g = g_0 \exp(-L/a_B)$. This gives $\beta = \ln(g/g_0)$. After connecting both asymptotic values by smooth lines, one obtains curves $\beta(\ln g)$ for different dimensions $d = 1, 2, 3$, as shown schematically in Fig. 1.4.

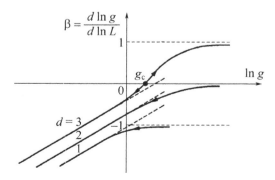

Fig. 1.4   Universal scaling functions $\beta(\ln g)$ for different dimensionality d. Arrows show the direction in which $\ln g$ varies when L increases. The point g_c corresponds to the critical 3d conductance, when the MIT occurs.

In the upper half-plane $\beta > 0$, the conductance increases with sample size L, which corresponds to a non-zero conductivity of a macroscopic sample at $T = 0$ (metal). In the lower half-plane $\beta < 0$, the conductance decreases exponentially with L and tends to zero for a macroscopic sample (insulator). Therefore, the MIT occurs upon crossing the point $\beta = 0$. One can see that the MIT exists for $d = 3$, whereas one-dimensional systems are always insulating at $T = 0$. The two-dimensional case is borderline. However, experience shows that for the two-dimensional electron gas, the conductivity also exhibits a transition to metallic conductivity with an increase of the electron density [2]. This is interpreted as a manifestation of the interactions in the electron system, which is not taken into account in the "scaling" model.

In 3d, $\beta = 0$ at the crossing point of the MIT, which means that the dimensionless conductance g_c and the real conductance G_c do not depend on the size of the sample L. As a result, the conductivity $\sigma \sim G_c/L$ goes continuously to zero with an increase in the sample size. In contrast to the concept of "minimal metallic conductivity" (Fig. 1.3(a)), the value $\sigma(0)$ increases continuously with $(x - x_c)$ as x exceeds the critical value x_c,

$$\sigma(0) \propto (x - x_c)^\mu. \tag{1.7}$$

For $x > x_c$, the critical exponent μ determines the dependence $\sigma(0)$ as a function of x (Fig. 1.3, curves b, c, d).

1.2   Impurity   Concentration   and   the   Metal–Insulator Transition

In 3d, most experimental studies were carried out for the conductivity of doped semiconductors with impurity concentration N as a parameter x in (1.7). Previous works show that the conductivity at zero temperature $\sigma(0) = \sigma(T \to 0)$ of doped semiconductors plotted as a function of impurity concentration N is equal to zero below some critical impurity concentration N_c. For $N > N_c$, $\sigma(0)$ obeys a power law in the vicinity of the transition

$$\sigma(0) \propto \left(\frac{N}{N_c} - 1\right)^\mu, \tag{1.8}$$

where μ is the critical conductivity exponent.

Most theoretical work [7–9] predicts $\mu = 1$, but there is also a prediction that $\mu = 1/2$ [10]. Numerical simulation [11] gives $\mu = 2/3$. A wide range of experimental values has been observed for μ with a distinction between uncompensated and compensated semiconductors. For uncompensated semiconductors (say, n-type), the concentration of electrons n is equal to the impurity concentration N_D (donors), while compensated semiconductors also have acceptor impurities of density N_A, therefore $n = N_D - N_A < N_D$.

An exponent μ close to unity has been found experimentally for most amorphous metal–insulator alloys [12, 13], Ge:Sb [14, 15], compensated Si:(P, B) [16], and Si:B in strong magnetic field [17]. However, values of μ close to 1/2 have been found for compensated Ge:As [18] and for all uncompensated silicon-based crystalline semiconductors, namely, $\mu = 0.65$ for Si:As and $\mu = 0.7$ for Si:(As, P) [19], $\mu = 0.5$ for Si:Sb [20], $\mu = 0.6$ for Si:As [21], $\mu = 0.65$ for Si:B in zero magnetic field B and $\mu = 0.8$ at $B = 1\,\text{T}$ [22]. Finally, $\mu = 0.5$ and 1.3 have been found in Si:P [23, 24]. Discussion of this discrepancy has been reviewed [25].

We see that in spite of many years of intense experimental and theoretical effort, the critical behavior of the conductivity in the vicinity of the MIT remains a puzzle.

The fact that different values of μ have been found in well characterized systems such as doped Si and Ge, leads one to consider carefully the question of the source of the discrepancy. Our results have been published in Ref. [26]. The conventional method for the determination of μ begins with the fitting of the measured conductivity data $\sigma(T)$ to the form $\sigma(T) = \sigma(0) + mT^p$ ($p = 1/2$ or $1/3$) associated with the quantum corrections to conductivity due to the electron–electron interaction [4] followed by the extrapolation of this dependence on the graph $\sigma(T)$ *vs.* T^p to the ordinate axis ($T = 0$). The accuracy of this method depends strongly on the low-temperature limit of the measured temperature interval and whether or not the dependence $\sigma(T)$ looks like a straight line on this scale. The determination of N_c is also given by the above procedure because, in accordance with its definition, the critical concentration is considered to be the maximal impurity concentration of the sample at which $\sigma(0) = 0$. If there is no such result in the series of samples investigated, researchers usually fit the data on the graph $\sigma(0)$ *vs.* N with a linear scale for all samples investigated and find the best agreement with Eq. (1.8) using N_c and μ as adjustable parameters.

The main source of inaccuracy of this method derives from the important contribution of those samples with N far from N_c in the fitting procedure, based on a linear scale, while the scaling theory is valid only in the immediate vicinity of N_c. Any inaccuracy in the determination of

N_c leads to an incorrect value of μ, because of the high sensitivity of the conventional method to small variations in N_c. For example, investigation of the same material Si:P by two groups [23] and [24] showed that a 6% decrease in N_c increased μ from 0.5 to 1.3, which demonstrates the very strong coupling between μ and N_c.

The obvious disadvantage of the conventional method results from the extrapolation of $\sigma(T)$ to $T = 0$. Strong fluctuations in the data at low temperatures, the influence of inhomogeneity of the samples, insufficient accuracy in the measurement of the Hall coefficient, the sharpness of the transition — all these factors become stronger upon approaching the critical point and on lowering the temperature.

We have suggested [26] a new approach for the determination of μ at nonzero temperatures. First, we pointed out that N_c can be obtained using an alternative idea. We proposed to include the fact that the temperature dependence of the conductivity $\sigma(T)$ for sample with concentration $N = N_c$ must obey $\sigma(T) = bT^{1/3}$ in accordance with the Aronov–Altshuler model [4]. It was shown that in the vicinity of the MIT, this law is observed for Ge:As and Ge:Sb. We introduce a new parameter $\alpha = [R(T_1)/R(T_2)] \cdot (T_1/T_2)^{1/3}$ where $R(T_1)$ and $R(T_2)$ are the resistances of a given sample measured at two temperatures in the interval where $\sigma(T) \sim T^{1/3}$ is observed. Then, the electron density in the sample with $\alpha = 1$ can be considered as a "critical concentration" N_c. The samples with $\alpha > 1$ should demonstrate metallic conductivity, while $\alpha < 1$ indicates the dielectric side of the transition if $T_1 > T_2$.

The principal point of our method is to consider the conductivity of the sample with $N = N_c$ as a ground level and replace $\sigma(0)$ by $\Delta\sigma(T^*) = \sigma_N(T^*) - \sigma_{N_c}(T^*)$ measured at any T^* where $\sigma(T) = a + bT^p$ ($p = 1/2$ or $1/3$). As a result, extrapolation to $T = 0$ is no longer needed.

The new method was applied to the investigation of the MIT in two series of low-compensated Ge samples doped by As and Sb. First, the value $\mu \sim 1$ was obtained using the conventional method, Eq. (1.8), based on the extrapolated values $\sigma(0)$. Then, $\mu = 1$ was obtained to high accuracy using our method based on replacing $\sigma(0)$ by $\Delta\sigma(T^*)$ in the form

$$\Delta\sigma(T^*) \propto \left(\frac{N}{N_c} - 1\right)^{\mu}. \tag{1.9}$$

It was shown that the value of μ obtained by the new method is insensitive both to the choice of the temperature T^* and to small variations of N_c. We also proposed to normalize $\Delta\sigma$ in order to compare the scaling behavior of the dimensionless ratio $\Delta\sigma/\tilde{\sigma}$ in different materials. We choose $\tilde{\sigma}$ of order the Mott "minimal metallic conductivity" $\tilde{\sigma} \approx \sigma_M = C_0(e^2/\hbar)N_c^{1/3}$ with C_0 as an adjustable numerical coefficient for a given semiconductor. We found that all the data for Ge:As and Ge:Sb merge together, taking into account only the difference in N_c. We also apply our method to those data obtained by other authors for Ge:Sb [15], Si:P [24], and Si:Sb [20]. We show that all the data for Si and Ge also merge into a universal dependence $\Delta\sigma/\tilde{\sigma} = \Delta N/N_c$ with $C_0 \approx 0.3$ for Ge and $C_0 \approx 0.6$ for Si.

Let us now discuss experimental results. Figure 1.5 shows the $\sigma(T)$ dependence in the $T^{1/3}$ scale. In the vicinity of the MIT, the curves correspond to $\sigma(T) \sim T^{1/3}$. We start by determining N_c. The temperatures $T_1 = 4.2\,\mathrm{K}$ and $T_2 = 2.0\,\mathrm{K}$ were chosen to calculate the parameter α. For three samples with $N = 3.58,\ 3.56,$ and $3.50 \times 10^{17}\,\mathrm{cm}^{-3}$, the values obtained for α are, respectively, 1.107, 1.071, and 0.999. The sample with $N = 3.50 \times 10^{17}\,\mathrm{cm}^{-3}$ has the value of α closest to unity. Therefore, this concentration can be taken to be the critical concentration N_c.

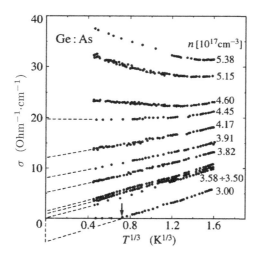

Fig. 1.5 Temperature dependence of conductivity in the vicinity of the MIT for a series of Ge:As samples plotted as a function of $T^{1/3}$ [26]. Impurity concentration in units of $10^{17}\,\mathrm{cm}^{-3}$ is shown near each curve.

The next step consists of taking the conductivity of this sample at any given temperature T^* as a ground level which allows the replacement $\sigma(0)$ by $\Delta\sigma(T^*) = \sigma_N(T^*) - \sigma_{N_c}(T^*)$. The low temperature, $T = 0.125\,\mathrm{K}$, was initially chosen for the calculation of $\Delta\sigma(T^*)$. One can see from Fig. 1.6 that the value of $\mu = 1$ is obtained without extrapolation to $T = 0$. The same dependence was plotted for two additional temperatures, $T_2^* = 0.512\,\text{K}$ and $T_3^* = 1\,\text{K}$. It was found that the data depend slightly on T^* for two reasons:

(i) in the immediate vicinity of the MIT, all curves $\sigma(T)$ are almost parallel (Fig. 1.5) and therefore $\Delta\sigma(T^*) \approx \Delta\sigma(0)$;

(ii) for more distant N, the temperature corrections to the value of conductivity are much smaller than those introduced by the increase of impurity concentration.

As a result, it was found that $\mu = 1$ for all temperatures. The same result was also obtained for another value of $N_c = 3.58 \times 10^{17}\,\text{cm}^{-3}$ of the neighboring sample. Correspondingly, all $\Delta\sigma(T^*)$ were recalculated taking into account the new ground level of conductivity.

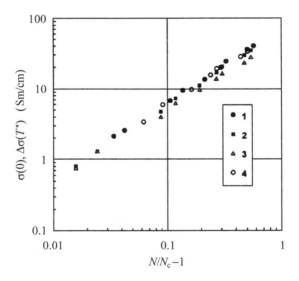

Fig. 1.6   Dependence $\sigma(0)$ (1) and $\Delta\sigma(T^*)$ (2–4) on the impurity concentration for the same series of Ge:As samples [26]. 1 — $N_c = 3.44 \times 10^{17}\,\text{cm}^{-3}$; 2 and 3 — $N_c = 3.50 \times 10^{17}\,\text{cm}^{-3}$, $T^* = 0.125$ and $1.0\,\text{K}$; 4 — $N_c = 3.58 \times 10^{17}\,\text{cm}^{-3}$, $T^* = 0.125\,\text{K}$.

This result is also presented in Fig. 1.6. The proposed method is insensitive to small variations in N_c because, in contrast with the values of $\sigma(0)$ obtained by the conventional method, our method leads to automatic corrections in the values of $\Delta\sigma(T^*)$ which compensate for the shift in N_c. Insensitivity to the small variation of N_c is an important advantage of the proposed method because it increases the reliability of the values obtained of μ. Similar results were obtained for Ge:Sb samples. Figure 1.7 shows that in the case of for Ge:Sb, $\sigma(T) \sim bT^{1/3}$ is observed at lower temperatures than in the case of Ge:As. For this reason, α was calculated for T_1^* and T_2^* below 1.0 K. The value $\alpha = 0.98$ was obtained for the sample with $N = 1.44 \times 10^{17}$ cm^{-3} which was taken as N_c for Ge:Sb. This value coincides with N_c obtained by the conventional method [15]. Determining μ from the concentration dependence of $\Delta\sigma(T^*)$, Eq. (1.9), using both our data and the data of Ref. [15], yields $\mu = 1$.

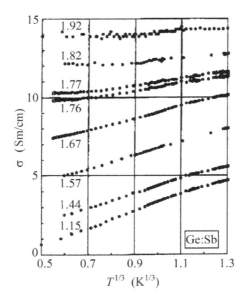

Fig. 1.7   Temperature dependence of conductivity for a series of Ge:Sb samples plotted as a function of $T^{1/3}$ [26]. Impurity concentration in units of 10^{17} cm^{-3} is shown near each curve.

Finally, we normalize $\Delta\sigma$ in order to compare the scaling behavior of the dimensionless ratio $\Delta\sigma/\tilde{\sigma}$ for Ge:As and Ge:Sb. As mentioned, we choose $\tilde{\sigma}$ of order the Mott minimum metallic conductivity $\tilde{\sigma} = C_0(e^2/\hbar)N_c^{1/3}$. Taking $C_0 \approx 0.3$ for Ge, ($\tilde{\sigma} = 55$ S/cm), all the data for Ge:As and Ge:Sb merge into a universal linear dependence $\Delta\sigma/\tilde{\sigma} = \Delta N/N_c$ (Fig. 1.8).

We show also the data obtained on the basis of the experimental results for Si:P [24] discussed above. Plotting their temperature dependence $\sigma(T)$ on a $T^{1/3}$ scale, we found $N_c = 3.55 \times 10^{18}$ cm^{-3}, which is close to the value $N_c = 3.52 \times 10^{18}$ cm^{-3} reported in [24]. The critical conductivity exponent determined by our method is $\mu = 1.0$ independent on the choice of T^* and small variations of N_c which significantly differs from the result $\mu = 1.3$ obtained in Ref. [24] by the conventional method. The normalized dependence $\Delta\sigma/\tilde{\sigma}$ (for Si we use $C_0 \approx 0.6$) is also shown in Fig. 1.8.

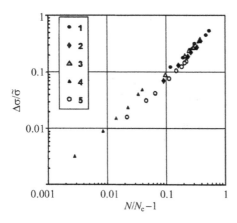

Fig. 1.8 Normalized conductivity $\Delta\sigma(T^*)/\tilde{\sigma}$ for different semiconductors as a function of $\Delta N/N_c$ [26]. The values of C_0 and N_c are given in the text. 1 — Ge:As, 2 — Ge:Sb, 3 — Ge:Sb, 4 — Si:P, 5 — Si:Sb.

It is interesting to note that there is a second special value of the impurity concentration in Fig. 1.5: at $N_M = 4.45 \times 10^{17}$ cm^{-3} for Ge:As, the derivative $\partial\sigma/\partial T$ changes sign at low temperatures. It is natural to believe that this is just the "minimal metallic conductivity", Eq. (1.4), because the classical and quantum corrections to the conductivity with opposite signs of the temperature dependence compensate each other at this concentration, and therefore σ became temperature-independent. At $N > N_M$, $\partial\sigma/\partial T$ is negative, which is typical for a "Drude metal". Substituting the value of $n = N_M$ in (1.4) gives $\sigma_M \approx 18$ S/cm, in good agreement with experiment. For Ge:Sb (Fig. 1.7), N_M is 1.92×10^{17} cm^{-3} which gives $\sigma_M \approx 14$ S/cm, also in agreement with experiment.

Figure 1.8 shows that all data for Ge, doped by As and Sb, and for Si, doped by P and Sb, merge into a universal scaling dependence with $\mu = 1$. Therefore, we may conclude that for all cases of uncompensated Ge:As, Ge:Sb, and Si:P, Si:Sb in the immediate vicinity of N_c, the critical conductivity exponent μ is equal to unity. Other values probably appeared as the result of attempts to extend the scaling procedure to N relatively far from the critical region. The question arises of why for compensated semiconductors, the conventional method gives the correct value of $\mu = 1$ even for samples with N relatively far from N_c. A possible explanation is that for compensated semiconductors, the critical region of the MIT is broadened because of the increasing amplitude of the random potential.

1.3 Magnetic Field Driven Metal–Insulator Transition. Universality of the Transition

We found, that for the case of the MIT driven by impurity concentration (N-MIT), the value of conductivity at zero temperature $\sigma(0) = \sigma(T \to 0)$ when plotted as a function of N is equal to zero on insulating side of the MIT and is finite above N_c, obeying the scaling law, Eq. (1.8) with critical exponent $\mu = 1$. For barely metallic samples with $N > N_c$, the MIT can occur upon the application of a critical magnetic field B_c because a strong magnetic field shrinks the electron wave function. Scaling behavior of conductivity is also expected in the neighborhood of a magnetic-field driven MIT (B-MIT):

$$\sigma(0) \propto \left(1 - \frac{B}{B_c}\right)^\nu.$$ (1.10)

where ν is also a critical exponent. Theoretical work [27, 28] predicts that $\nu = \mu = 1$.

The B-MIT in the form Eq. (1.10) has been described [29–31] and ν was found to be close to unity. We tried to combine the two descriptions of the MIT using the same series of samples Ge:As for which the N-MIT is shown in Fig. 1.5. At $B = 0$, $N_c = 3.5 \times 10^{17} \, \text{cm}^{-3}$. Figure 1.9 shows B-MIT for one of the samples with $N = 4.6 \times 10^{17} \, \text{cm}^{-3}$, which is above N_c.

In accordance with Fig. 1.5, this sample is in the metallic region. However, it can be converted into an insulator by applying a magnetic field which in this case plays the role of the control parameter x. One can see from Fig. 1.9 that for B-MIT, similar to N-MIT, $\sigma(T) = a + bT^{1/3}$ and crosses the transition at a critical magnetic field B_c where $a = 0$. For this sample $B_c = 5 \, \text{T}$. For stronger magnetic fields, 7 and 8 T, the experimental points deviate from the straight line at very low temperatures, due to hopping conductivity (similar bending in Fig. 1.5 looks like a break).

To determine the critical index ν, we plot $\sigma(0) = \sigma(T \to 0)$ as a function of the scaling variable $[1 - (B/B_c)]$. This dependence is shown in the inset (a) to Fig. 1.10. To avoid the inaccuracy connected with extrapolation to $T = 0$, we have used the same method as for N-MIT: we consider $\sigma(T)$ of any sample in critical regime (i.e., at $B = B_c$) as a ground level of conductivity, and instead of $\sigma(0)$, we plot the measured quantity $\Delta\sigma(T^*) =$

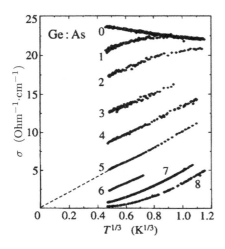

Fig. 1.9 Temperature dependence of conductivity of a sample of Ge:As with $N = 4.6 \times 10^{17}$ cm^{-3} in different magnetic fields [31]. Magnetic field in Tesla is shown near each curve.

$\sigma_{\rm B}(T^*) - \sigma_{B_{\rm c}}(T^*)$ at temperature T^* within the interval of T where $\sigma(T) \propto bT^{1/3}$ is observed. The data for $T^* = 0.1$ K and 0.216 K are shown in the inset to Fig. 1.10.

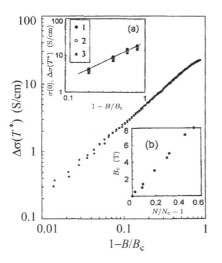

Fig. 1.10 Dependence of $\Delta\sigma(T^*)$ at $T^* = 0.216$ K as a function of $[1 - (B/B_{\rm c})]$ for one sample of Ge:As. Inset (a) shows the scaling dependence of $\sigma(0) - 1$, $\Delta\sigma(T^*)$ measured at $T^* = 0.1$ K $- 2$, and at 0.216 K $- 3$ for the sample of Ge:As. The straight line corresponds to $\nu = 1$. Inset (b) shows the dependence of $B_{\rm c}$ on $\Delta N/N_{\rm c}$ for series of Ge:As samples.

One can see that the resulting value $\nu = 1$ does not depend on this replacement. The main panel of Fig. 1.10 shows $\Delta\sigma(T^*)$ at $T^* = 0.216$ K for the sample with $N = 4.6 \times 10^{17}$ cm^{-3}, as a function of $[1 - (B/B_{\rm c})]$ with $B_{\rm c} = 5$ T. Measurements of $B_{\rm c}$ for different samples show that $B_{\rm c}$

increases linearly with increasing $\Delta N \equiv N - N_c$:

$$B_c = B_0 \frac{\Delta N}{N_c}. \tag{1.11}$$

This dependence is shown in the inset (b) to Fig. 1.10. Here $B_0 = 15\,T$ for the series of Ge:As samples. The linear dependence allows us to combine of both N-MIT and B-MIT by introducing a new scaling variable $U = [(N/N_c - 1)(1 - B/B_c)]$, or, by taking Eq. (1.11) into account,

$$U = \frac{N}{N_c} - \frac{B}{B_0} - 1. \tag{1.12}$$

Knowledge of B_0 and N_c allows one to calculate U for any sample with $N > N_c$ and $B < B_c$ and plot $\sigma(0)$ or $\Delta\sigma(T^*)$ as a function of U. To normalize the conductivity, we plot the dimensionless ratio $\sigma(0)/\tilde\sigma$ or $\Delta\sigma(T^*)/\tilde\sigma$, $\tilde\sigma = C_0(e^2/\hbar)N_c^{1/3}$ with $C_0 = 0.3$ for doped Ge, as in Fig. 1.8. The result is shown in Fig. 1.11. All the data exhibit universal scaling behavior

$$\frac{\sigma(0)}{\tilde\sigma} = U^\mu, \qquad \mu = 1. \tag{1.13}$$

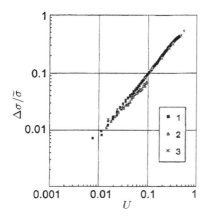

Fig. 1.11 Normalized scaling conductivity $\sigma(0)/\tilde\sigma$ for Ge:As *vs.* new universal variable $U = (N/N_c) - (B/B_0) - 1$ [31]. 1, 2 — B-MIT for two samples of Ge:As with $N = 5.38$ and 4.17×10^{17} cm^{-3} ($N_c = 3.50 \times 10^{17}$ cm^{-3}, $B_0 = 15\,T$), 3 — N-MIT for series of Ge:As at $B = 0$.

Thus, we see that the MIT induced by variation of either impurity concentration N or magnetic fields B is continuous with critical exponent $\mu = 1$.

1.4 Temperature-Driven Metal–Insulator Transition

The MIT in the conductivity is a manifestation of a transition of the Fermi level (FL) from delocalized to localized electron states. For the case of doped semiconductors, this means that the DC (direct current) conductivity at zero temperature $\sigma(0)$ when plotted as a function of impurity concentration N is equal to zero on insulating side of the MIT and remains finite on the metallic side, at $N > N_c$, obeying the power law (1.8) in the vicinity of the transition (N-MIT). At non-zero temperatures, the low-temperature conductivity $\sigma(T)$ is also determined by the position of the FL in the density-of-states (DOS). If the FL is situated in the localized part of the DOS, $\sigma(T)$ is governed by hopping conductivity which decreases exponentially with decreasing T:

$$\sigma(T) = \sigma_0 \exp\left(-\frac{\varepsilon_h}{kT}\right), \tag{1.14}$$

where ε_h is the energy of activation for hopping conductivity (see Chapter 2). By contrast, the conductivity of electrons on the metallic side, in delocalized states, could be characterized by non-exponential, power law behavior $\sigma(T) = \sigma(0) + mT^p$, where $p = 1/3$ in the immediate vicinity of the MIT [1]:

$$\sigma(T) = a + bT^{1/3}. \tag{1.15}$$

Thus, at non-zero temperatures, one can define the transition from localized to delocalized electron states as a transition from exponential (1.14) to non-exponential (1.15) temperature dependence of the DC conductivity.

We now consider the temperature-induced transition from localized to delocalized states in barely insulating samples of Ge:As and Ge:Sb with N below the critical concentration N_c of the MIT [32, 33]. We show that the conductivity of these samples exhibits metallic-like behavior above some "delocalization temperature" T_d. This means that the concentration boundary between samples with insulating and metallic behavior of conductivity, $N_c(T)$, decreases with increasing temperature. We will refer to this effect as temperature-induced metal–insulator transition (T-MIT). We show that the values of dimensionless temperature T_d/W, where $W = (e^2/\kappa)N_c^{1/3}$ is the Coulomb energy on the mean inter-impurity distance (κ is the dielectric constant of the impurity host semiconductor), merge for different impurity systems into an universal "phase diagram" of the transition.

The N-MIT in investigated samples was described in the previous paragraph. It was shown that for Ge:As, $N_c = 3.50 \times 10^{17}\,\mathrm{cm}^{-3}$, for Ge:Sb, $N_c = 1.44 \times 10^{17}\,\mathrm{cm}^{-3}$ and the critical conductivity exponent $\mu = 1$ for both series of samples. Figures 1.5 and 1.7 show the $\sigma(T)$ dependence for both series of samples as a function of $T^{1/3}$. For barely metallic samples with $N > N_c$, the $T^{1/3}$-dependence of $\sigma(T)$ is closely represented by a straight line, in agreement with (1.15). On the other hand, one can see in Fig. 1.5, that the conductivity of the barely insulating sample with $N = 3.00 \times 10^{17}$ cm^{-3} also demonstrates metallic-like behavior $\sigma(T) \propto T^{1/3}$ at $T > T_{\mathrm{d}}$, where T_{d}, indicated by an arrow in Fig. 1.5, could be called as a "delocalization temperature". This means that samples with N just below N_{c} can be considered as insulating only at $T = 0$ and at very low temperatures. However, if T increases above $T_{\mathrm{d}}(N)$, these samples exhibit metallic-like behavior for the conductivity. In other words, the value of N_{c}, which separates insulating and metallic behavior for the conductivity, decreases with increasing T.

To obtain the temperature dependence of N_{c}, we plot the conductivity at different temperatures as a function of N in the dimensionless scale $\sigma/\tilde{\sigma}$ vs. $N/N_{\mathrm{c}} - 1$, where $\tilde{\sigma}$ is of order the Mott minimum metallic conductivity. Figure 1.12 shows the results for a series of Ge:As samples.

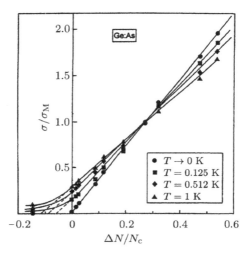

Fig. 1.12 The normalized conductivity for series of Ge:As samples at different temperatures as a function of the dimensionless impurity concentration [32].

The values of $\sigma(0)$ for all samples were determined from Fig. 1.5 by extrapolating $\sigma(T)$ to $T = 0$. Figure 1.12 shows that the linear character of the scaling behavior of $\sigma(0)$ persists to non-zero temperatures as well. Extrapolation of these straight lines to the X-axis gives the values of $N_{\mathrm{c}}(T)$. We next plot T_{d} as a function of $[N_{\mathrm{c}}(T)/N_{\mathrm{c}}(0) - 1]$. A similar procedure has been carried out for a series of Ge:Sb samples and for the data for Si:P and

Si:B published in Refs. [24] and [17]. The results are shown in Fig. 1.13.

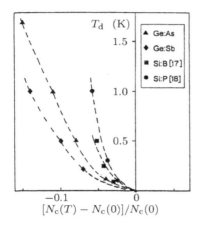

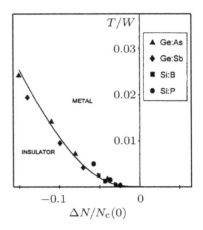

Fig. 1.13 Dependence of the "delocalization temperature" T_d on impurity concentration for different semiconductor systems [32].

Fig. 1.14 The "phase diagram" of the MIT: dependence of critical concentration of the transition on the temperature in normalize units [32, 33].

The "delocalization temperature" T_d is seen to increase rapidly with decreasing N below $N_c(0)$. Moreover, this effect is stronger for materials with higher N_c. Therefore, the energy of Coulomb interaction on the mean inter-impurity distance $W = (e^2/\kappa)N_c^{1/3}$ was chosen in an attempt to normalize the "delocalization temperature". The mean energy of the random Coulomb potential $W = (e^2/\kappa_0)N_{c0}^{1/3}$ is 70 K for Ge:As, 52 K for Ge:Sb, 206 K for Si:B and 200 K for Si:P.

The result of normalization is presented in Fig. 1.14. The solid line corresponds to the parabolic dependence $y \sim x^2$, where $y = T/W$ and $x = \Delta N_c/N_c(0)$. All data merge into one curve, which can be considered a "phase diagram" of the metal–insulator transition.

For barely insulating samples at non-zero temperatures, the critical concentration of the MIT is shifted to lower concentration. It is valid, of course, only within the narrow interval of concentrations and temperatures.

We conclude this chapter with the claim that the "exponent puzzle" has now been solved: $\mu = 1$ in different kinds of the MIT. This claim is supported in a review entitled "Fifty years of Anderson localization" [34, 35].

Chapter 2

Different Types of Hopping Conductivity

2.1 Basic Concepts of Hopping Mechanism of Electron Transport

There is a number of books devoted to hopping conductivity (see, for example, Refs. [1–3]) to which we refer the reader who wants to get acquainted with the problem in more detail. Here the main concepts of the hopping mechanism of transport will be briefly described.

At non-zero temperature, localized electrons may contribute to the electron transport due to hopping of charge carriers between one localized states to another. Usually hopping conductivity in crystalline semiconductors dominates at low temperatures when free carriers with high mobility are frozen out. In many amorphous and insulating materials, where the concentration of free carriers is negligible, while the density of defects, impurities and other trapping centers for electrons is very high, hopping is the main mechanism of conductivity up to room temperatures and even higher. However, in what follows, we will consider an impurity band in doped semiconductors which is a good model for the study of hopping conductivity.

Important point of theoretical consideration is an assumption that all localized states have different energy: two centers with the same energy are separated by the infinite distance, so the density-of-states (DOS) in impurity band is of statistical origin, which is fundamentally different from the DOS in the conduction band, where delocalized states belong to each point of the crystal. In Fig. 2.1 the energy-level scheme and density of localized states are shown in an impurity band in n-type semiconductors for the case of weak Fig. 2.1(a) and strong Fig. 2.1(b) degree of compensation $K = N_A/N_D$ which indicate the ratio between minority and majority impurities (acceptors and donors for n-type semiconductors). Compensation

is a mandatory condition for hopping of electrons from an occupied donor
to empty one, because existence of empty places at $T = 0$ is provided by
$K \neq 0$.

In Fig. 2.1, short dashes show the level of donors, and circles show
electrons localized on some of them. E_c is the bottom of the conduction
band, E_0 is the ionization energy for a single isolated donor, μ is the position
of the Fermi level. The density-of-states is shown to the right. The occupied
states are shaded.

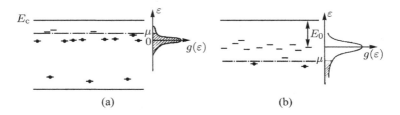

Fig. 2.1   Schematic representation of the energy and space distribution of the localized
states in impurity band in n-type semiconductor in the case of weak (a) and strong (b)
compensation.

First model of hopping conductivity has been suggested by Miller and
Abrahams (MA) on the following approach [4]. Starting with electron wave
functions localized on individual donors, calculate the probability that an
electron transition will occur between two donors i and j with the emission
or absorption of a phonon. Then calculate the number of transitions γ_{ij} per
unit time. In the absence of an electric field, an equal number of electrons
undergo the reverse transition, i.e., there is a detailed balance. In a weak
electric field the forward and reverse transitions will not be balanced, giving
rise to a current proportional to the field. Evaluating this current yields
the resistance R_{ij} of a given transition, and thus the whole problem is
reduced to calculating the electrical conductivity of an equivalent network
of random resistors.

Let us briefly present the main results of this approach. We denote the
wave function of electron localized at the point r_i, $\Psi(r - r_j)$, as Ψ_i, with
the average energy $\langle \varepsilon_i \rangle$ and the wave function of electron localized at the
point r_j, $\Psi(r - r_j)$, as Ψ_j, with the average energy $\langle \varepsilon_j \rangle$ (Fig. 2.2).

The averaging reflects the fact that energy of each level fluctuates in time
due to fluctuating Coulomb potential connected with continuously changed

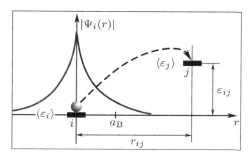

Fig. 2.2 Schematic representation of an average energy and position for an occupied donor i and an empty donor j.

distribution of other electrons among the localized centers (impurities). It is assumed that the distance between localized centers $r_{ij} = r_j - r_i$ is much larger than the radius of localization a. Therefore, probability for electron to transfer the space distance r_{ij} is proportional to the overlap integral which is proportional to $\exp(-2r_{ij}/a)$. The transfer needs absorption of a resonance phonon with energy $\varepsilon_{ij} = |\langle\varepsilon_i\rangle - \langle\varepsilon_j\rangle|$. Probability to find the phonon with energy ε_{ij} is determined by the Plank distribution function

$$P(\varepsilon) = \left[\exp\left(\frac{\varepsilon}{kT}\right) - 1\right]^{-1}, \tag{2.1}$$

which at low temperatures ($\varepsilon_{ij} \gg kT$), gives the second small factor $\exp(-\varepsilon_{ij}/(kT))$ in the probability of the electron transfer between states i and j:

$$\gamma_{ij} = \gamma_0 \exp\left(-\frac{2r_{ij}}{a}\right)\exp\left(-\frac{\varepsilon_{ij}}{kT}\right). \tag{2.2}$$

Finally, the transition can take place only if the initial state i is occupied and the final state j is empty. Both probabilities are determined by the Fermi distribution function

$$f(\varepsilon) = \left[\exp\left(\frac{\varepsilon - \mu}{kT}\right) + 1\right]^{-1}. \tag{2.3}$$

Here μ is the energy of the Fermi level. Let $n = (0, 1)$ be the n'-th donor occupation number which fluctuates in time. The transition $i \to j$ is possible only when $n_i = 1$ and $n_j = 0$. Therefore the number of electrons making this transition per unit time is given by

$$\Gamma_{ij} = \langle\gamma_{ij}n_i(1 - n_j)\rangle, \tag{2.4}$$

where the averaging is over time. The quantity Γ_{ij} also fluctuates in time. This happens because of the fluctuating in occupation numbers for donors neighboring i and j. Variations with time in the potential of these donors gives rise to fluctuations in n_i, n_j, and hence in Γ_{ij}.

In further theoretical consideration, a very important simplifying assumption has been made. It is assumed that the site occupation numbers and energies do not fluctuate in time, but remain equal to their average values. This approximation consists in the following:

(i) Each donor is characterized by the average occupation number $\langle n_i \rangle = f_i$.

(ii) Corresponding to each donor, there is a time-averaged electronic level energy in the field of all other impurities and electrons:

$$\varepsilon_i = \sum_{1}^{\text{acc}} \frac{e^2}{\kappa |r_i - r_l|} - \sum_{k \neq i}^{\text{don}} \frac{e^2(1 - f_k)}{\kappa |r_i - r_k|}. \tag{2.5}$$

Here the first sum extends over all acceptors, and the second — over all donors except i; the quantity $e(1 - f_k)$ represents the average charge of donor k.

(iii) The phonon energy absorbed in the transition $i \to j$ is taken to be $\varepsilon_{ij} = \varepsilon_j - \varepsilon_i$. In this approximation

$$\Gamma_{ij} = \gamma_0 \exp\left(-\frac{2r_{ij}}{a}\right) P(\varepsilon_{ij}) f_i(1 - f_j). \tag{2.6}$$

For the reverse process, $j \to i$, with the emission of a phonon, we have, by analogy:

$$\Gamma_{ji} = \gamma_0 \exp\left(-\frac{2r_{ij}}{a}\right) \left[P(\varepsilon_{ij}) + 1\right] f_j(1 - f_i). \tag{2.7}$$

We can write the current between the donors i and j in the form:

$$J_{ij} = -e(\Gamma_{ij} - \Gamma_{ji}).$$

In the absence of an electric field E, the functions f_i are given by the equilibrium expression (2.3). Substituting Eq. (2.3) into Eqs. (2.6) and (2.7), one obtain that in equilibrium there is a detailed balance between the transitions $i \to j$ and $j \to i$; at $E = 0$, $\Gamma_{ij} = \Gamma_{ji} = \Gamma_{ij}^0$ and hence also $J_{ij} = 0$. So, the frequency Γ_{ij}^0 of the transitions $i \to j$ and $j \to i$ in equilibrium has the form

$$\Gamma_{ji}^0 = \gamma_0 \exp\left(-\frac{2r_{ij}}{a}\right) \exp\left(-\frac{\varepsilon_{ij}}{kT}\right),$$
$$\varepsilon_{ij} = \frac{1}{2}\left(|\varepsilon_i - \varepsilon_j| + |\varepsilon_i - \mu| + |\varepsilon_j - \mu|\right). \tag{2.8}$$

Here ε_{ij} reflects both the energy difference between states i and j and the probability to find an occupied initial state and an empty final state.

This balance is, of course, destroyed by an electric field. Firstly, the field redistributes electrons over donors, creating a shift of the Fermi level μ and corrections to distribution function $f_i(E)$ and affects the donor-level energies ε_i via both direct action of the external field E and variation in the Coulomb potential due to a redistribution of electrons. This gives rise to a change in the absorbed phonon energy $\varepsilon_i - \varepsilon_j$ which enters the argument of the Planck function. One can express these corrections by introducing small quantities $\delta\mu_i$ and $\delta\varepsilon_i$. If the electric field is so small that the corrections $\delta\mu_i$ and $\delta\varepsilon_j$ are small compared to kT, then one can expand the functions as power series in these corrections. In an approximation which is linear in the external field, one find

$$J_{ij} = \frac{e}{kT}\, \Gamma_{ij}^0 \left[\delta\mu_j + \delta\varepsilon_j - (\delta\mu_i + \delta\varepsilon_i) \right],$$

This equation could be rewritten in the form of the Ohm's law:

$$J_{ij} = \frac{U_i - U_j}{R_{ij}}, \qquad (2.9)$$

where

$$R_{ij} = \frac{kT}{e^2 \Gamma_{ij}^0}; \qquad (2.10)$$

and $eU_i = \delta\mu_i + \delta\varepsilon_i$ is a local value of the electrochemical potential on donor i, counted from the electron chemical potential μ. Therefore $U_i - U_j$ can be interpreted as a voltage drop on the transition $i \to j$, and R_{ij} as the resistance of this transition.

Thus the conductivity of the sample is completely determined by the resistances R_{ij} which are proportional to the product of two exponents:

$$R_{ij} = R_{ij}^0 \exp(\xi_{ij}), \qquad \xi_{ij} = \frac{2r_{ij}}{a} + \frac{\varepsilon_{ij}}{kT}. \qquad (2.11)$$

The hopping conductivity problem is reduced to calculation of the conductivity of a random Miller–Abraham (MA) network which has its vertices at the donors and in which resistance R_{ij} connects each pair of vertices (Fig. 2.3).

An important feature of the resistance network shown in Fig. 2.3, is its extremely wide spectrum of resistances R_{ij}. Calculation of the MA network is based on "percolation" principle: the greatest resistances are removed from the network sequentially as long as the network remains simply connected. After removal, the dense of the initial MA network

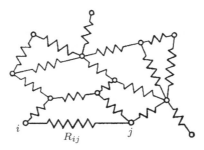

Fig. 2.3   Miller and Abrahams' random resistance network [4].

decreases continuously and resistance of the whole network depends on the greatest resistance $R_c = R_0 \exp(\xi_c)$ which one had to leave without loss of connectivity. The value of $\xi = \xi_c$ of the resistance R_c is called "percolation threshold". Therefore, hopping resistivity can be written as

$$\rho = \rho_0 \exp(\xi_c), \qquad \xi_c = \frac{2r_c}{a} + \frac{\varepsilon_c}{kT}, \qquad (2.12)$$

where a prefactor ρ_0 contains information about the host semiconductor material, like deformation potential, crystal density, speed of sound, constant of the electron–phonon interaction, etc.

As a result, the current flows through the part of donors which form so-called "infinite cluster" which is passing through the whole sample.

Figure 2.4 shows schematically the "infinite cluster" together with large "finite" clusters inside the cells and "dead ends".

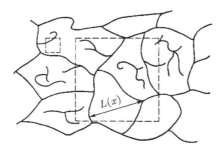

Fig. 2.4   Schematic representation of the "infinite" cluster [1]. The lines are actually chains of donors. L is the correlation length which represents the average size of the cells. The dashed squares have sides $l > L$ and $l < L$. Large "finite" clusters inside cells and "dead ends" are also shown.

The lines are actually chains of donors with resistances between them $R_{ij} \leq$

R_c. The correlation radius L corresponds to the average size of the cells. If one include also larger resistances with $\xi = \xi_c + 1$, the infinite cluster become more dense, but the larger resistances will be shortened by the existing path and will not change significantly the total conductivity.

2.2 Nearest Neighbor Hopping Conductivity

The simplest kind of hopping conductivity is nearest neighbor hopping (NNH). The density-of-states (DOS) in the impurity band has a maximum at the energy of about energy of ionization for an isolated donor E_0 (Fig. 2.1). In the case of weak or strong compensation, the DOS at the Fermi level is small, therefore the average distance between these states is large, much larger than the mean distance between donors $r_D \sim N_D^{-1/3}$. It is naturally to assume that if the initial and final states of hopping are nearest neighbors, they have energy in the vicinity of the maximum of DOS which is located at the level E_0. In the case of strong compensation shown in Fig. 2.1(b), almost all states at the level E_0 are empty, and conductivity is proportional to the probability for electrons to be ionized from the Fermi level to the level $E_0 : \exp(-|\mu - E_0|/(kT))$. For small compensation, the Fermi level is above E_0, and almost all states in the vicinity of E_0 are occupied (Fig. 2.1(a)). To realize the hopping, the final state must be empty. The probability of such event depends on the energy distance between the Fermi level μ and maximum of the DOS, which is also proportional to $\exp(-|\mu - E_0|/(kT))$. All other energies ε_{ij} between donors in the vicinity of E_0 which appear during the hopping motion, are much less than $|\mu - E_0|$ and therefore $\varepsilon_c = |\mu - E_0|$ is the main parameter in Eq. (2.13) which determines NNH.

The critical percolation distance r_c is close to the mean separation between donors r_D with concentration N, $r_D = (4\pi N_D/3)^{-1/3}$, so $r_c \approx r_D \propto (N_D)^{-1/3}$. This leads to the following expression for NNH (in terms of resistivity):

$$\rho(T) = \rho_0 \exp\left(\frac{C_3}{N^{1/3}a}\right) \exp\left(\frac{\varepsilon_3}{kT}\right), \qquad \varepsilon_3 = |\mu - E_0|. \qquad (2.13)$$

Numerical coefficient $C_3 = 1.73$ was calculated on the basis of the percolation approach (number "3" is traditionally attributed to NNH). It was shown theoretically that for slightly doped and weakly compensated semiconductors, the value of ε_3 is proportional to the energy of Coulomb interaction on the mean distance between impurities $\varepsilon_D = e^2/(\kappa r_D)$, or more precisely [1],

$$\varepsilon_3 = 0.61\,\varepsilon_D = 0.99\frac{e^2}{\kappa}N_D^{1/3}. \tag{2.14}$$

Meaning of the coefficient 0.99 is purely symbolic — to show that the Eq. (2.14) is obtained as a result of rigorous calculation, and is not "the order of unity". The dependence (2.14) has been verified experimentally. Figure 2.5 shows the dependence of ε_3 on the concentration of different majority impurities in Ge at weak compensation (see Ref. [5] and references therein).

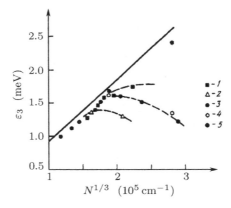

Fig. 2.5 Dependence of the energy of activation ε_3 on concentration of majority impurities in Ge at low degree of compensation. 1 — series of samples with P, 2 — samples with Ga, 3 — samples with Sb, 4 — two samples with Sb, 5 — the same two samples with Sb subjected to uniaxial stress along $\langle 111 \rangle$ axis. Solid line represents the theoretical dependence (2.15) [5].

One can see that, indeed, at small impurity concentration, experimental points agree with theoretical dependence (2.14), but at some concentration $N_D = N_{\max}$, the dependence $\varepsilon_3(N_D^{1/3})$ has maximum and then goes down slowly to zero.

This phenomenon could be naturally explained by an increase of overlapping the donor wave function. In this case the correction to the classical dependence (2.14) will contain the Plank constant and Bohr radius a. Dependence on a is seen in Fig. 2.5. There is a relationship between Bohr radius for impurities Ga, Sb and P in Ge: $a(\mathrm{Ga}) > a(\mathrm{Sb}) > a(\mathrm{P})$. Therefore, the effect of overlapping has to be observed in the same sequence, which is in agreement with experiment. The assumption about the quantum reason for the deviation of ε_3 from the dependence (2.14) can be seen more clearly from the experiment with uniaxial stress [5]. In Ge, the uniaxial stress along $\langle 111 \rangle$ direction lifts the 4-fold valley degeneracy and strongly decreases the overlap of the wave functions. Therefore, if the deviation is determined by overlapping, the stress has to decrease this deviation.

Figure 2.6 shows that it is exactly what happened in the experiment.

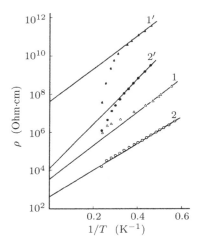

Fig. 2.6 Temperature dependence of the resistivity for two samples Ge:Sb without stress (1, 2) and under the maximal uniaxial stress along $\langle 111 \rangle$ direction (1′, 2′).

For more doped sample 2 ($N_{Sb} = 2.2 \times 10^{16}\,\mathrm{cm^{-3}}$, $N^{1/3} = 2.8 \times 10^5\,\mathrm{cm^{-1}}$), the value of ε_3 increased significantly under uniaxial stress and almost reached the predicted theoretical value (2.14). At the same time, for the sample 1 with lower doping ($N_{Sb} = 6.5 \times 10^{15}\,\mathrm{cm^{-3}}$, $N^{1/3} = 1.9 \times 10^5\,\mathrm{cm^{-1}}$), in which ε_3 is close to theoretical value, the uniaxial stress does not change ε_3. The values of ε_3 for stressed samples are plotted in Fig. 2.5.

For slightly and moderately doped semiconductors, $r_D/a \gg 1$, i.e. $N^{1/3}a \ll 1$, therefore, at relatively high T, the first exponential term in Eq. (2.13) is much larger than the second one. With decrease of T, the second term increases, but until it is smaller than the first one, the energy of activation $\varepsilon_h = \varepsilon_3$ remains constant. However, if temperature continues to decrease, the second term in Eq. (2.13) became as large as the first one. This occurs at temperature

$$T_c \approx \frac{1}{k}\frac{\varepsilon^2}{\kappa}N_D^{2/3}a. \tag{2.15}$$

To minimize the resistivity below this T_c, hopping will be realized between states with energy in closer vicinity of the Fermi level μ where obviously there exist occupied and empty places. This leads to the new kind of hopping conductivity called Variable-Range-Hopping (VRH).

2.3 Variable-Range-Hopping Conductivity; the "Mott Law"

The idea of VRH and the below considerations were put forward first by N. F. Mott [6]. Assume that the density-of-states (DOS) in the vicinity of Fermi level (FL) μ is constant $g(\varepsilon) = g(\mu)$. Let us consider an energy band of the width ε around the FL: $\mu \pm \varepsilon/2$. The number of states in this vicinity is $N(\varepsilon) = g(\mu)\varepsilon$, the average distance between them is $r(\varepsilon) \approx [N(\varepsilon)]^{-1/3} = [g(\mu)\varepsilon]^{-1/3}$. For the main percolation parameter ξ_c, in Eq. (2.12) one obtain the following expression (from here we measure energy and temperature in the same units):

$$\xi_c = \frac{2}{[g(\mu)\varepsilon]^{1/3}a} + \frac{\varepsilon}{T}. \tag{2.16}$$

Maximal conductivity will be realized at minimal ξ_c when $d\xi_c/d\varepsilon = 0$. This gives

$$\varepsilon = \left[\frac{T}{g(\mu)^{1/3}a}\right]^{3/4} = (T^3 \cdot T_M)^{1/4}, \qquad T_M = [g(\mu)a^3]^{-1}. \tag{2.17}$$

Correspondingly,

$$r(T) = a\left(\frac{T_M}{T}\right)^{1/4}. \tag{2.18}$$

Numerical calculation based on percolation approach [1] gives for three-dimensional conductivity the value $T_M = C_M[g(\mu)a^3]^{-1}$, where $C_M = 21.2$. One can see that energy of activation continuously decreases with temperature as $\varepsilon \sim T^{3/4}$, while the average hopping distance increases, $r \sim T^{-1/4}$. This is why this mechanism is called Variable-Range-Hopping (VRH).

Substitution Eqs. (2.17) and (2.18) in Eq. (2.12) gives the so called the "Mott law" or "$T^{-1/4}$ law":

$$\rho(T) = \rho_0 \exp\left(\frac{T_M}{T}\right)^{1/4}, \qquad T_M = 21.2[g(\mu)a^3]^{-1}. \tag{2.19}$$

It is easy to expand this law for other dimensionality d. In this case, DOS is d-dimensional and the power $1/3$ in Eq. (2.16) must be replaced by $1/d$. Then we can repeat the previous arguments and get Eq. (2.19) in the form

$$\rho(T) = \rho_0 \exp\left(\frac{T_M}{T}\right)^{1/(d+1)}.$$

Particularly, for 2d conductivity

$$\rho(T) = \rho_0 \exp\left(\frac{T_M}{T}\right)^{1/3}, \qquad T_M = 13.8[g(\mu)a^2]^{-1}. \qquad (2.20)$$

Thus, at relatively high temperatures, hopping conductivity is carried out by means of NNH mechanism with constant energy of activation $\varepsilon_3 = |\mu - E_0|$. With decrease of T, the value of ε_3/T continuously increases and come up to 1.73 $(N^{1/3}a)^{-1}$, Eq. (2.14). This leads to the transition from NNH to VRH conductivity which manifests itself in continuous decrease of the energy of activation with decrease of temperature.

2.4 Coulomb Gap in the Density-of-states; the "Efros–Shklovskii Law"

In the previous consideration, it was assumed that DOS in the vicinity of the Fermi level is constant. However, taking into account that the energy of each donor site ε_i depends on distribution of other charged donors and acceptors, Eq. (2.5), one can conclude that due to the Coulomb interaction, electrons localized on states near the Fermi level (FL) will be redistributed to reduce the total electron energy. This leads to formation of a "soft" Coulomb gap at FL with $g(\mu) = 0$.

Let us consider two states, i and j from the energy interval $\mu - \varepsilon/2$ and $\mu + \varepsilon/2$, below and above the Fermi level μ. This means that in equilibrium, at $T = 0$, state i with $\varepsilon_i < \mu$ is occupied by an electron, and state j with $\varepsilon_j > \mu$ is empty. These states are separated by a distance r_{ij}. What happens if electron jumps from center i to center j? First, we remove electron from site i to infinity, where potential energy is equal to zero. Then the state i will be positively charged and the energy level ε_j decreases down to $\varepsilon_j - e^2/(\kappa r_{ij})$ (κ is the permittivity of the host semiconductor). However, the new energy value of state j must remain higher than ε_i, as otherwise, a new configuration will be preferable when, conversely, state j is occupied and state i is empty in equilibrium. From inequalities

$$\varepsilon_j - \varepsilon_i - \frac{e^2}{\kappa r_{ij}} > 0, \qquad \varepsilon_j - \varepsilon_i < \varepsilon \qquad (2.21)$$

we obtain the following limitation for ε from below and for the number of states $N(\varepsilon)$ from above:

$$\varepsilon > \frac{e^2}{\kappa r_{ij}}, \qquad N(\varepsilon) \approx (r_{ij})^{-3} < \left(\frac{\kappa \varepsilon}{e^2}\right)^3. \qquad (2.22)$$

Since the energy ε can be arbitrary small, we can, following Ref. [3], replace the sign of inequality for equality and after differentiation we obtain for DOS:

$$g(\varepsilon) = \frac{\partial N}{\partial \varepsilon} \approx (\varepsilon - \mu)^2 \left(\frac{\kappa}{e^2}\right)^3. \tag{2.23}$$

This parabolic form of the DOS is called "soft gap" because $g(\varepsilon) = 0$ only at one point $\varepsilon = \mu$.

For two-dimensional (2d) conductivity, similar consideration leads to the linear soft gap:

$$g(\varepsilon) \sim |\varepsilon - \mu| \left(\frac{\kappa}{e^2}\right)^2. \tag{2.24}$$

It follows from (2.22), that $r_{ij} = e^2/(\kappa\varepsilon)$ and therefore the mean hopping distance r_c increases with decrease of the energy interval ε around the Fermi level regardless of the dimension. As a result, we obtain the following expression for the main percolation parameter ξ_c:

$$\xi_c = \frac{2e^2}{\kappa a \varepsilon} + \frac{\varepsilon}{T} \tag{2.25}$$

with minimum ($d\xi_c/d\varepsilon = 0$) at

$$\varepsilon = \left(T\frac{e^2}{\kappa a}\right)^{1/2} = (T \cdot T_{ES})^{1/2}, \qquad T_{ES} \approx \frac{e^2}{\kappa a}. \tag{2.26}$$

This leads to the "Efros–Shklovskii" (ES) law (or "$T^{-1/2}$" law) for VRH conductivity (in terms of resistivity):

$$\rho(T) = \rho_0 \exp\left(\frac{T_{ES}}{T}\right)^{1/2}, \qquad T_{ES} = 2.8\frac{e^2}{\kappa a}. \tag{2.27}$$

Thus, the Coulomb interaction between localized carries leads to appearance of a soft gap at the Fermi level, which is parabolic in the case of 3d, Eq. (2.23), Fig. 2.7.

Denote the half-width of this gap as Δ. At relatively high temperatures when $T > \Delta$, the Mott law (or "$T^{-1/4}$" law) has to be observed, because the Coulomb gap does not influence very much the integral density-of-states far from the FL. With decrease of T, when $T < \Delta$, the Mott law must be replaced by the ES law which leads to a transition from "$T^{-1/4}$" to "$T^{-1/2}$" law.

So, with decreasing temperature, three kinds of hopping conductivity should be observed in following consequence: NNR, VRH of the Mott type and VRH of the ES type. Experimental observations of both transitions: from NNH to the Mott VRH and from the Mott law to the ES law are described below.

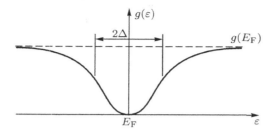

Fig. 2.7   Schematics of the 3d density-of-states in the vicinity of the Fermi level.

2.5   Transition from Nearest Neighbor to Variable-Range-Hopping Conductivity

Hopping conductivity with constant energy of activation has been often observed in many doped semiconductors, including classical materials like Ge, Si, GaAs. The main feature which justifies the hopping origin of electron transport is very strong dependence of the value of conductivity (or resistivity) on the concentration of impurities, Eq. (2.13). The most cited is Fig. 2.8, adopted from Ref. [7] where the temperature dependence of resistivity is shown for series of samples of p-Ge in very wide interval of impurity concentration and constant degree of compensation $K = 40$ %.

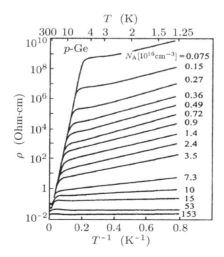

Fig. 2.8   Resistivity of samples of neutron-transmutation-doped (NTD) p-Ge with constant degree of compensation $K = 0.4$ and different concentration of impurities [7].

One can see that increasing the impurity concentration by a factor of 30 results in a value of conductivity which is of 7 orders of magnitude lower!

Small technical insert. The important feature of this experiment is the method of doping which provides high level of homogeneity and absence of any correlation in the impurity distribution. This method is based on irradiation of pure Ge samples with thermal neutrons flow in a nuclear reactor and is called neutron-transmutation doping (NTD) [8, 9]. In NTD, three of five stable isotopes of Ge are transmuted after capturing of a thermal neutron and emission of a γ-quant ((n, γ)-reaction):

$$^{70}\text{Ge}(n,\gamma)^{71}\text{Ge} \rightarrow {}^{71}\text{Ga, shallow acceptor,}$$
$$^{74}\text{Ge}(n,\gamma)^{75}\text{Ge} \rightarrow {}^{75}\text{As, shallow donor,} \qquad (2.28)$$
$$^{76}\text{Ge}(n,\gamma)^{77}\text{Ge} \rightarrow {}^{77}\text{As} \rightarrow {}^{77}\text{Se, deep donor.}$$

The ratio of the concentration of the different impurities is determined by the thermal-neutron cross section and the abundance of the given isotopes. For natural Ge, the NTD leads to creation of p-type Ge with 40 % compensation, the concentration of impurities linearly depends on the time of irradiation and can be controlled with high accuracy. Due to small value of the thermal-neutron cross section, the doping events are distributed uniformly across the sample. These two features (the precisely controlled level of doping and uniform impurity distribution) make the NTD method very useful for investigation of conductivity in the vicinity of MIT.

One can see also from Fig. 2.8 that in the "liquid Helium" interval of temperatures (from 4.2 K down to 1.3 K), the energy of activation ε_3 for each sample is constant which is characteristic for NNH.

To observe the transition from NNH to VRH, it is necessary to reduce the temperature below 1 K. For two samples from the series of NTD-Ge, these measurements were performed and the results are described in Ref. [10].

Small technical insert. In experiments described in Ref. [10], the temperature below 1 K was obtained by adiabatic demagnetization of paramagnetic potassium chrome alum $\text{CrK(SO}_4)_2 12\,\text{H}_2\text{O}$. The temperature was determined by measuring the magnetic susceptibility of the salt. Two methods were used to produce thermal contact between the sample and the block of paramagnetic salt. In the first, the salt block was made in the form of a cylinder with an axial hole. The sample was placed freely, without stress, inside the cylinder, and four wires (current and potential probes) were glued to the salt, thus ensuring good thermal contact at $T > 0.1$ K. In the second case, the sample was placed in a container with a water–glycerine solution of the alum. The solution solidified when cooled, causing hydrostatic compression of the sample. In this case, thermal contact between the salt and the sample was sufficiently reliable at $T > 70$ mK.

The temperature dependence of the resistivity of the measured sample in the interval $4.2 \div 1.3$ K is shown in Fig. 2.9(a). It is seen that hydrostatic compression (the dark circles) leads to an increase of ε_3 and also made it possible to transform two samples with metallic conductivity to insulating state with $\varepsilon_3 \neq 0$. Thus, the use of the second method of mounting of the sample resulted in a larger assortment of curves with the available samples.

Figure 2.9(b) shows the same plots of $\lg \rho$ vs. $1/T$ down to 100 mK. It is seen that ε_3 does not remain constant but decreases continuously with decreasing temperature. This is a clear evidence of the VRH mechanism.

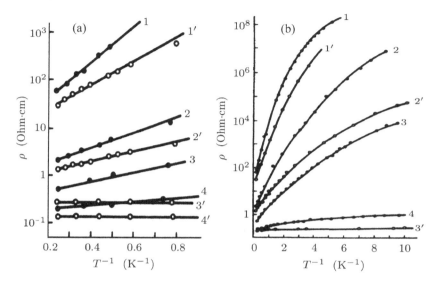

Fig. 2.9 Temperature dependence of resistivity of four NTD p-Ge samples without stress (open circles) and with hydrostatic stress (dark circles). (a) — in the narrow interval of "liquid helium" temperatures $(4.2 \div 1.3 \text{ K})$ and (b)— in wider temperature range (down to 0.1 K)

It is interesting to notice that the curves have a smooth form starting with 4.2 K, so that apparent linearity in the $4.2 \div 1.3$ K interval is due to the narrowness of the temperature interval. This means that straight lines in Fig. 2.8 do not represent the constant ε_3, as claimed in many books and reviews. This misunderstanding is explained by the form of the DOS $g(\varepsilon)$ in the case of intermediate degree of compensation $K \approx 0.5$. For weak $(K \ll 1)$ or strong $(1 - K \ll 1)$ compensation, DOS at the FL $g(\mu)$ is much less than that one at the level E_0, where DOS reaches its maximum value (see Fig. 2.1).

That is why conductivity has a constant energy of activation $\varepsilon_3 = |\mu - E_0|$ in a relatively extended temperature range. In the case of $K \approx 0.5$, the FL is located just near the maximum of the DOS which gives no range for the constant energy of activation. Even if we take into account the Coulomb gap at the Fermi level, where $g(\mu) = 0$, the situation is not changed significantly because in this case, the width of the Coulomb gap 2Δ is approximately equal to the width of the impurity band. Figure 2.10 shows the results of computer modeling the DOS at $K = 0.5$ with and without Coulomb gap [11].

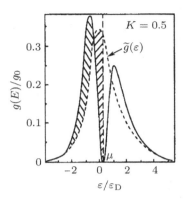

Fig. 2.10 Computer modelling of the density of states in impurity band for the case of degree of compensation $K = 0.5$ with and without taking into account the Coulomb interaction between localized carriers [11]. The shaded areas are equal.

2.6   Transition from the Mott Law to the Efros–Shklovskii Law

As we have seen, the VRH conductivity can be observed in two forms, as Mott VRH and as Efros–Shklovskii (ES) VRH. In general form, one can write the temperature dependence of hopping conductivity as

$$\sigma(T) = \sigma_0 \exp\left(-\frac{T_p}{T}\right)^p, \qquad (2.29)$$

where both parameters T_p and power exponent p depend on the DOS near the FL $g(E_F)$. Since cooling reduces continuously the width of the "optimal" energy band E_c containing the states that participate in the hopping process, $p < 1$. If the DOS is constant or varies slowly, $g(\varepsilon) = g(E_F)$ and VRH conductivity is described by the Mott law:

$$p = \frac{1}{4}, \qquad T_{1/4} \equiv T_M = 21[g(E_F)a^3]^{-1}. \qquad (2.30)$$

If the parabolic (in 3d) Coulomb gap appears at the Fermi level

$$g(E_F) = 0, \qquad g(\varepsilon) = g_0(\varepsilon - E_F)^2 \equiv g_0\varepsilon^2. \qquad (2.31)$$

Here energy is measured from the Fermi level, and

$$g_0 = \frac{3}{\pi}\left(\frac{e^2}{\kappa}\right)^{-3}. \qquad (2.32)$$

In this case, VRH conductivity is described by the ES law:

$$p = \frac{1}{2}, \qquad T_{1/2} \equiv T_{\mathrm{ES}} = 2.8\left(\frac{e^2}{\kappa a}\right). \tag{2.33}$$

There are many published observations of the VRH conductivity which obeys laws with both $p = 1/2$ and $1/4$ (see, for example, Ref. [2]). A transition from the ES to the Mott law was observed in the same sample with an increase in the temperature [12, 13]. As mentioned, such a transition is explained by the fact that the Coulomb interaction can alter the DOS only near the FL. Far from this level the DOS is restored to its initial value, which is approximately equal to $g(E_{\mathrm{F}})$ (Fig. 2.7). The half-width of the Coulomb gap Δ can therefore, be determined from $g_0\Delta^2 = g(E_{\mathrm{F}})$ which gives

$$\Delta = \left[\frac{g(E_{\mathrm{F}})}{g_0}\right]^{1/2}. \tag{2.34}$$

At $T \ll \Delta$, the ES law is obeyed ($p = 1/2$). In the opposite case ($T \gg \Delta$), $p = 1/4$ and VRH has to be obeyed the Mott law.

Lets estimate the crossover temperature T_c. At high T, when $T \gg \Delta$, the width of the optimal bend $\varepsilon(T)$ of localized levels involved in VRH is $\varepsilon(T) = T^{3/4}[g(E_{\mathrm{F}})a^3]^{-1/4}$, see (2.16). It is naturally to suggest that at the crossover temperature, $\varepsilon(T_c) = \Delta$, where Δ is the width of the Coulomb gap, (2.34). This gives

$$T_c = \left(\frac{e^2}{\kappa}\right)^2 g(E_{\mathrm{F}})a. \tag{2.35}$$

Taking into account expressions for T_{M} (2.30) and T_{ES} (2.33), the Eq. (2.35) may be rewritten in terms of measurable parameters:

$$T_c = 2.7\frac{T_{\mathrm{ES}}^2}{T_{\mathrm{M}}}. \tag{2.36}$$

Let's turn to experimental data. The described below crossover between ES and Mott VRH conductivity was studied in weakly compensated n-Ge:As samples [13].

Small technical insert. The samples were prepared from single crystals of germanium, doped by the shallow donor impurity arsenide. In order to measure the resistivity of a semiconductor at very low T, the concentration of impurities n must be very close to the Mott concentration n_c at which a metal–insulator transition takes place. Because we are quite near the metal–insulator transition, the control upon the level of doping must be carried out to high accuracy. The described above NTD method, Eq. (2.28), provides such an accuracy, but for natural Ge, NTD technique leads to creation of a p-type material with 40 % compensation. In order to obtain n-type material with small compensation, a specially grown crystal of Ge was used, enriched artificially with the isotope ^{74}Ge up to 98 %. As a result, a series of homogeneously doped samples of n-Ge:As was obtained. One of the samples has an impurity concentration of $n = 3.2 \times 10^{17}$ cm^{-3}, which is close to the critical concentration of 3.5×10^{17} cm^{-3} for Ge:As (see Chapter 1). Since $n = 0.9n_c$, it allows us to prolong the measurements of the resistance down to 30 mK.

Figure 2.11 shows the temperature dependence of the sample resistance $R(T)$ plotted together in two scales: $\lg R$ *vs.* $T^{-1/2}$ and *vs.* $T^{-1/4}$.

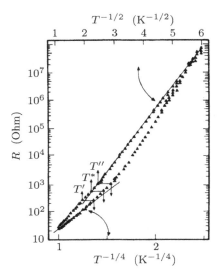

Fig. 2.11   Temperature dependence of the n-Ge:As sample resistance plotted in $\lg R$ *vs.* $T^{-1/4}$ and $T^{-1/2}$ scales [13].

One can see that the low-temperature part of $R(T)$ below $T'' = 0.17$ K is better straightened in the scale $T^{-1/2}$ which corresponds to the ES law. The deviation from the straight line in the scale "$T^{-1/2}$" with increase of T is interpreted as a crossover to the Mott law. One can see that, indeed, at $T > T' = 0.31$ K, and up to 1 K, the curve $R(T)$ is better straightened in the scale "$T^{-1/4}$" (lower scale in Fig. 2.11).

If we assume that there is a real "$T^{-1/4}$" law, we can estimate T_c and compare it with experiment. Thus, in Fig. 2.11, the experimental "transition interval" from "$T^{-1/4}$" law to a "$T^{-1/2}$" law is between $T' = 0.31$ K and $T'' = 0.17$ K with a midpoint of $T_c = 1/2(T' + T'') = 0.24$ K which is indicated as T^*. From the slopes of straight lines in "$T^{-1/4}$" and "$T^{-1/2}$" scales, we obtain $T_M = 1400 \pm 350$ K and $T_{ES} = 10.5 \pm 1.5$ K. Substituting these values to Eq. (2.36) we calculate $T_c = 0.21 \pm 0.04$ K in good agreement with experimental value $T^* = 0.24$ K. This confirms correctness of numerical coefficients in theoretical expressions for T_M and T_{ES}.

2.7 Non-Ohmic Hopping Conductivity

Until now, we have discussed the temperature dependence of hopping conductivity in the ohmic regime, when the resistance of the sample does not depend on the applied electric field. A weak electric field leads to only a slight redistribution of electrons over donors in an impurity band, which in turn caused a small changes in the distribution function $\delta\mu_i$ and the donor-level energies $\delta\varepsilon_i$, see Eqs. (2.9–2.11). Investigations of the hopping transport in strong electric fields in the case of an impurity band were limited by the small energy of ionization for shallow impurities, or, in other words, due to nearness of impurity band to the conducting band. In this case, increase of an electric field leads to rapid ionization of impurities, so the conductivity of delocalized electrons ceases to be of hopping nature. For the same reason, VRH conductivity in the case of small and intermediate degree of compensation ($K \ll 1$, $K \approx 0.5$), is observed, as we have seen, at very low temperatures, below 1 K.

However, in the case of strong compensation, only the deepest impurity levels are occupied by electrons, so the Fermi level μ moves into the depths of the impurity band (Fig. 2.1(b)). Therefore, it is better to study the influence of electric fields on hopping conductivity in heavily doped ($N^{1/3}a \geq 1$) and compensated ($1 - K \ll 1$) semiconductors (HDC). It has been shown in Ref. [14] that VRH conductivity in HDC-Ge is observed at $T > 1$ K, in so-called "liquid helium" interval of temperatures (4.2 ÷ 1.3 K). In this experimental conditions, samples are immersed in liquid helium which allows to dissipate more power and therefore allows to apply higher voltage and current in contrast with experiments below 1 K in dilution refrigerator, when samples are in vacuum and cooling is realized only through the very thin probe wires.

The influence of an electric field on VRH conductivity was studied in two samples of HDC n-Ge [15] (concentration of phosphorous donor impurity $N = 10^{18}$ cm^{-3}, compensation $K \approx 0.8$). In weak electric fields, the temperature dependence of resistivity of these samples in the "liquid helium" temperature interval is obeyed the Mott law of VRH (Fig. 2.12):

$$\rho(T) = \rho_0 \exp\left(\frac{T_0}{T}\right)^{1/4}, \qquad T_0 = \frac{21.4}{g(\mu)a^3}. \qquad (2.37)$$

The values of T_0 determined from the slope of straight lines are 1.1×10^6 K and 6.6×10^5 K for samples 1 and 2 correspondingly. However, measurements of the current–voltage ($I - V$) characteristics showed that the ohmic regime of conductivity for these samples is narrow and decreases

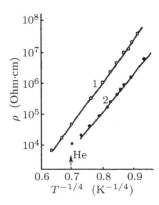

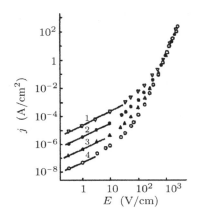

Fig. 2.12   Temperature dependence of resistivity of samples 1 and 2 HDC-Ge [15].

Fig. 2.13   Current–voltage characteristics of the sample 1 HDC-Ge [15]. T, K: 1 — 4.2, 2 — 3.0, 3 — 2.2, 4 — 1.6.

with decrease of temperature. Figure 2.13 shows the $I - V$ characteristics of one of the HDC-Ge sample at different temperatures plotted in units of the current density $j\,(\mathrm{A/cm^2})$ *vs.* electric field E (V/cm) in logarithmic scale. One can distinguish three areas:

(i) weak fields, where j linearly increases with E (ohmic behavior, shown as straight lines), in this area resistivity $\rho = E/j$ does not depend on E and depends only on T;

(ii) strong fields, where j does not depend on T and depends only on E;

(iii) intermediate fields, where j depends both on T and on E. Let us discuss all these areas.

2.7.1   *Weak electric fields*

The upper limit of the ohmic regime is explained by the fact that electric field decreases the energy difference between initial and final states by the value eEr, if an electron hops on the distance r along the field. Of course, in the "isotropic infinite cluster" (Fig. 2.4), the hopping transitions occur in different directions, but the main contribution in the increase of current belongs to the hopping along the field. Therefore, one can assume that the ohmic regime will be valid until $eEr < T$, which gives for the upper limit the value $E_c = T/(er)$. In the NNH conductivity, when $r = r_c \approx N^{-1/3}$ is constant, E_c decreases linearly with T. In the VRH conductivity $r \approx$

$a(T_0/T)^{1/4}$ increases with decrease of T as $T^{-1/4}$, Eq. (2.18), where a is
the Bohr radius. Increase of r leads to an additional decrease of E_c with T.
One can see from Fig. 2.13, that, indeed, E_c rapidly decreases with decrease
of temperature, so the field interval of ohmic regime quickly narrows. The
known values of T_0 allow us to determine the ratio r/a and then, using
$a = 4$ nm for P impurity in Ge [1], one can obtain r. This gives for $r(T)$
the values from 35 nm at 4.2 K to 65 nm at 1.6 K for both samples.

2.7.2 *Strong electric fields*

In HDC-Ge, the energy distance between Fermi level and allowed band is
relatively large, so all the effects associated with the transfer of carriers
from localized states to the conducting band (like Poole–Frenkel effect)
can be ignored and therefore, the temperature independence of the current
density can be considered as a manifestation of an activationless hopping
conductivity. The idea of this mechanism was first discussed by Mott [16].
As is known, in weak electric fields, when the slope of Fermi level can
be ignored, the hopping process involves phonon absorption and emission
with equal frequency. The absorption of phonons is just responsible for
the exponential temperature dependence of hopping conductivity. Mott
pointed out that in a strong electric field, when the fall of potential energy of
an electron $eEr(T)$ on the typical length of jump $r(T)$ becomes comparable
with the width of a band of energies around the Fermi level $\varepsilon(T)$, electrons
can move along the field direction without absorption of phonons, but only
emitting them at every jump. For this case, the field dependence of the
current density was obtained in Ref. [17]:

$$j \sim \exp\left[-\left(\frac{E_0}{E}\right)^{1/4}\right], \qquad E_0 = \gamma\frac{T_0}{ea}, \qquad (2.38)$$

where T_0 is the same as in Eq. (2.37), γ is a numerical coefficient of order
of 1.

Derivation of Eq. (2.38) is the same as in the case of Mott law. Requirement for an electron to be transferred from the filled to empty state
located at distance r "down the field" without absorption of a phonon
means that the energy difference between these states does not exceed eEr.
Therefore, the effective integral density of states involved in the hopping, is
$N_{\text{eff}}(r) = g(\mu)eEr$ and average distance of hopping is $r \approx (N_{\text{eff}})^{-1/3}$, which
gives $r(E) \approx [eEg(\mu)]^{-1/4}$. After substituting $r(E)$ into the expression for
probability for electron to jump over a distance r: $w(r) \sim \exp(-2r/a)$ we
obtain Eq. (2.38).

Figure 2.14 shows the dependences $\lg j$ *vs.* $E^{-1/4}$ in the area of strong fields for the same two samples of HDC Ge as in Fig. 2.13.

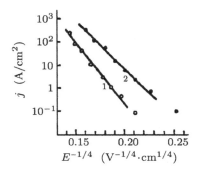

Fig. 2.14 Dependence of $\lg j$ *vs.* $E^{-1/4}$ for samples 1 and 2 HDC-Ge at $T = 1.6$ K [15].

The values E_0 are 2.4×10^8 and 1.1×10^8 V/cm for samples 1 and 2 correspondingly. Using E_0 and T_0 obtained above from the low field data, we can estimate from Eq. (2.38) the values of a (with accuracy of the factor of γ). We obtain 4 nm and 5 nm which are, indeed, close to the known value of the Bohr radius for P donor in Ge.

2.7.3 *Intermediate electric fields*

If electric fields are strong enough, one can take into account basically only the hopes of electrons in the direction of field, so, in Eq. (2.12) one can change ε_c to $\varepsilon_c - eEr$. For Mott VRH, this leads to similar temperature dependence like $\exp[(T_0/T^*)^{1/4}]$, if we assume that T^* is now the function of E. At $eEa < T$, this dependence is simplified to $T^* = T[1 + (\beta eEa)/T]$, where β is a numerical coefficient which is determined by the fact that hops are realized not within the sphere with the center in which electron is initially localized, but within the volume elongated along the field. Under the circumstances, conductivity has to be obeyed the relation [18, 19]:

$$\ln \frac{\sigma(E)}{\sigma(0)} = \frac{eEl}{T}, \qquad (2.39)$$

where $l = r(T)$ is a length parameter equal to the hopping length $r(T)$. In Ref. [20], the topology of the "infinite cluster" was involved into consideration and a conclusion was made that l in Eq. (2.39) must be equal to the correlation radius of the "infinite cluster" $L = r(T)[\xi_c/(\xi - \xi_c)]^\nu \gg r(T)$ (see Fig. 2.4). Here $\nu = 0.9$ and ξ_c is the critical percolation parameter, Eq. (2.12).

Experiment allowed us to verify these two predictions. One can see in Fig. 2.15, that the dependence $\lg[\sigma(E)/\sigma(0)]$ is indeed a linear function of the applied field.

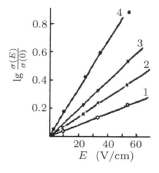

Fig. 2.15 Field dependence of $\lg[\sigma(E)/\sigma(0)]$ for sample 1 [15]. T, K: 1 — 4.2, 2 — 3.0, 3 — 2.2, 4 — 1.6.

The slopes of straight lines in Fig. 2.15 give the length $l = 35$ nm at 4.2 K to 60 nm at 1.6 K, in good agreement with the values of $r(T)$, obtained from the temperature dependence of resistivity in the ohmic regime.

This fact proves the correctness of the model suggested in Refs. [18, 19] and contradicts to the prediction made in Ref. [20], may be due to deformation of the "infinite cluster" in strong electric fields. In reality, increasing of E above E_c leads to "straightening" of the current lines and deformation of the infinite cluster, it becomes elongated in the direction of electric field (Fig. 2.16).

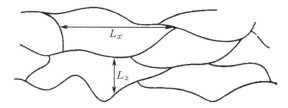

Fig. 2.16 Schematic representation of a critical subnetwork for the case of strong electric field [1]. Line indicate resistance chains. "Dead ends" are not shown.

If in weak fields, the optimal states for hopping are distributed with equal probability at any angle to the direction of field, then in strong fields, the increase of conductivity is determined by the decrease of the energy level only for states located in the narrow solid angle around the field direction. Straightening of the current line leads also to decrease of the number of hops required to pass between electrodes.

Evaluating the dependence of $r(T, E)$ in the whole range of electric field, one can mention that in strong fields, the value of r decreases similar to that with increase of temperature, see Eqs. (2.38) and (2.37). In this sense, one can say about "heating up" of localized electrons by electric field, although it is obvious that localized electrons cannot take energy directly from an electric field during the hopping. In this case, "heating up" is described by the non-equilibrium distribution function, which depends on E.

2.7.4 Negative differential hopping conductivity

As we have seen (Fig. 2.13) at $E > E_c$, increase of electric field leads to significant increase of hopping conductivity. This behavior could be called "positive hopping electroconductivity". However, in some experiments the opposite effect was found, when hopping conductivity decreases with increase of an applied electric field, a "negative electroconductivity" (NEC) [21, 22]. The NEC effect was observed in lightly doped samples with the nearest neighbor hopping (NNH) regime with constant energy of activation ε_3 and in a narrow interval of temperatures, when all localized charge carriers are already excited from the Fermi level μ to the level E_0 of maximal DOS (Fig. 2.1(a)), but ionization to the conductivity band (ε_1-conductivity) still not begins. As a result, a "plateau" in the temperature dependence of hopping resistivity is observed (see the upper curves in Fig. 2.8, temperature interval between 3 K and 5–7 K).

In Ref. [21], experiments were performed on series of five lightly doped n-Ge:P samples with small degree of compensation. Concentration of phosphorous donor impurity was within the interval from 3.7 to 10.5×10^{15} cm^{-3}. It was shown that $\rho(T)$ for all of these samples obeyed the NNH mechanism with constant ε_3 [5].

Field dependences of conductivity of one of the samples n-Ge:P with $N_P = 6.0 \times 10^{15}$ cm^{-3} (sample 3) at different temperatures are shown in Fig. 2.17.

Non-ohmic part of the curves shows unexpected effect of negative "electroconductivity" (NEC) at relatively weak fields and low temperatures: conductivity slightly decreases with increase of E and then sharply increases. NEC is observed only in a narrow interval of impurity concentration N (Fig. 2.18).

The explanation of NEC was suggested in Ref. [23]. The model is based on existence of the "dead ends" in the texture of the infinite cluster (see

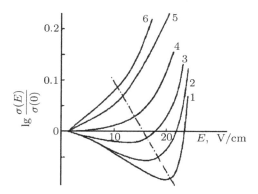

Fig. 2.17 Negative electroconductivity for sample 3 Ge:P [21]. T, K: 1 — 4.2, 2 — 2.5, 3 — 2.25, 4 — 2.0, 5 — 1.63, 6 — 1.44.

Fig. 2.4). Fragment of the infinity cluster with dead ends is schematically shown in Fig. 2.19. In the ohmic regime ($E < E_c$), the dead ends does not influence the conductivity because electrons can hope in and out of the dead end. However, with increase of E, the jumps in the opposite to the field direction become increasingly difficult, and therefore some of the dead ends act as a trapping center. The number of field-induced traps increases with E. This leads to a decrease of concentration of freely hopping charge carriers which is maximal in the temperature interval corresponding to saturation of NNH. In its turn, this leads to decrease of hopping conductivity with increase of electric field.

If concentration of charge carriers, excited to the level E_0, is not maximal (lower temperatures), electric field positively influences the conductivity, facilitating the hops due to decrease of the energy difference between hopping states. This positive effect is stronger than the negative, which explains why the NEC is not observed with decrease of T (Fig. 2.17). Similar arguments can be used for explanation of the optimal interval of impurity concentration where the NEC effect is observed (Fig. 2.18).

One can see that with the further increase of E, NEC is sharply changed to the positive effect. This sharp transition is explained by the topology of the infinite cluster. The last impurity center in the dead end is separated from another center of the infinite cluster by distance $r_c + \Delta r > r_c$ (Fig. 2.19).

With increase of E, the probability for electron to go out from the dead end against the field decreases as $\exp(-eEX/T)$, where X is projection of the dead end length on the field direction. At the same time, increase of E increases the probability for an electron to escape the dead end along

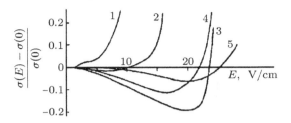

Fig. 2.18   Dependences $\Delta\sigma(E)/\sigma(0)$ at $T = 4.2$ K for series of samples n-Ge:P [21].
$N(10^{15}\text{cm}^{-3})$: 1 — 3.7, 2 — 4.5, 3 — 6.0, 4 — 7.4, 5 — 10.5.

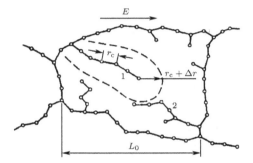

Fig. 2.19   Fragment of an infinite cluster with dead ends [22]. The dashed line shows the insulating area surrounding the electric field-induced electron trapping in the dead end.

the field, overcoming an additional small factor $\exp(-2\Delta r/a)$ for this hop. Therefore the maximal field E_m which can trap electron in the dead end of length X is determined from $eE_m X/T = 2\Delta r/a$. Further increase of E leads to the sharp "breakdown" of the dead end and "delocalization" of the trapped electrons for freely hopping.

So, we see that the unexpected effect of negative electroconductivity (NEC) in the hopping regime is understood qualitatively. Quantitative analysis requires a more precise knowledge of the topology and structure of the "infinite cluster".

Chapter 3

"Hopping Spectroscopy"

In the VRH regime the temperature dependence of resistivity can be written as

$$\rho(T) = \rho_0 \exp\left(\frac{T_p}{T}\right)^p, \tag{3.1}$$

where the energy parameter T_p and the exponent p depend on the form of the density of localized states (DOS) $g(\varepsilon)$ near the Fermi level $g(E_F)$: in the case of constant or slowly varying $g(\varepsilon)$, $p = 1/4$ for three-dimensional conductivity and $p = 1/3$ for two-dimensional conductivity (Mott law), while in the case of existence of a soft Coulomb gap (CG) around the Fermi level with $g(E_F) = 0$, $p = 1/2$ for both 3d and 2d VRH (Efros–Shklovskii, ES law).

In this chapter we show how the measurements of the resistance and magnetoresistance in the VRH regime could be used for detection of the DOS in the vicinity of the Fermi level.

3.1 Three-dimensional Variable-Range-Hopping

In contrast to the nearest-neighbor hopping, the VRH mechanism involves only localized states in the vicinity of the Fermi level, the so-called "optimal band". The lower the temperature, the narrower is this "optimal band" whose half-width E_c is given by $E_c \sim T^{1-p} = T\xi_c$, where $\xi_c = (T_0/T)^p$ is the critical percolation parameter, Eq. (2.12). The value of $\xi_c = \ln[\rho(T)/\rho_0]$ can be obtained from experimental data. In the case of 3d, the mean hopping distance $r(T) = [I(E_c)]^{-1/3}$ where $I(E_c)$ is the integrated density of localized states in the "optimal band":

$$I(E)_c = \int_{-E_c}^{E_c} g(\varepsilon)d\varepsilon = 2\int_0^{E_c} g(\varepsilon)d\varepsilon. \tag{3.2}$$

For a constant DOS at the Fermi level $g(E_F)$, $I(E_c)$ is a linear function of E_c:

$$I(E_c) = 2g(E_F)E_c. \tag{3.3a}$$

In the opposite case, within the Coulomb gap (CG), one get $g(\varepsilon) = g_0\varepsilon^2$ for 3d VRH, which leads to a cubic dependence

$$I(E_c) = \frac{2}{3}g_0E_c^3. \tag{3.3b}$$

Usually, the ES law ($p = 1/2$) corresponds to the experimental data only at the lowest temperatures, for which $E_c \ll \Delta$, where Δ is the half-width of the CG, $\Delta = [g(E_F)/g_0]^{1/2}$ (see Eq. (2.34)). The entire temperature range can be accurately described by a model which allows for a smooth crossover from the ES law to the Mott law, i.e., within a model which takes into account the actual spectrum of the density of localized states. In this analysis we need to know the profile of the DOS (see Fig. 2.7). The expression for $g(\varepsilon)$ should have two limiting forms: $g(\varepsilon) = g_0\varepsilon^2$ for $\varepsilon \ll \Delta$ and $g(\varepsilon) = g(E_F)$ for $\varepsilon \gg \Delta$. One can suggest the following extended function for $g(\varepsilon)$ [1, 2]:

$$g(\varepsilon) = \frac{g_0\varepsilon^2}{1 + (\varepsilon/\Delta)^2}. \tag{3.4}$$

Substituting Eq. (3.4) into Eq. (3.2) yields

$$I(E_c) = 2g_0\Delta^3\Big[\frac{E_c}{\Delta} - \arctan\Big(\frac{E_c}{\Delta}\Big)\Big], \tag{3.5}$$

which gives the following limiting forms:

$$I(E_c) = \frac{2}{3}g_0E_c^3, \qquad E_c \ll \Delta; \tag{3.6}$$

$$I(E_c) = 2g_0\Delta^3\Big(\frac{E_c}{\Delta} - \frac{\pi}{2}\Big), \qquad E_c \gg \Delta. \tag{3.7}$$

One can see that Eq. (3.6) is equal to Eq. (3.3b) and that Eq. (3.7) is equal to Eq. (3.3a) minus a term $\pi g_0\Delta^3$ because of the existence the CG near the Fermi level. The general expression (3.5) exhibits an extended crossover phenomenon: a smooth transition from a cubic to a linear dependence of $I(E_c)$.

It follows from Eq. (3.4) that to determine the DOS, it is necessary to know two parameters, g_0 and Δ. In turn, g_0 is governed by the value of the permittivity κ, $g_0 = (3/\pi)(e^2/k)^{-3}$, see Eq. (2.32), which for the case of proximity of the electron concentration n to the critical concentration n_c of the metal–insulator transition may differ considerably from the value κ_0 for lightly doped semiconductors (for Ge, $\kappa_0 = 16$). At the lowest temperatures, when the ES law is valid, the value of κ could be obtained from the temperature dependence of the resistance $R(T)$ plotted as $\ln R$ *vs.* $T^{-1/2}$. From the slope of the straight line one can determine $T_{1/2} = 2.8(e^2/(\kappa a))$. However, one first needs to find independently the localization radius a, which differs from the Bohr radius a_B for individual donor impurities due to the proximity of to the metal–insulator transition.

The procedure for determining a is described in Refs. [3] and [4]. It is based on measurements of the hopping magnetoresistance (MR) in weak magnetic fields B. Theoretical considerations show [5] that in the hopping regime, MR is positive due to the shrinkage of the electron wave function and must be proportional to $\exp(B^2)$:

$$\ln \frac{R(B)}{R(0)} = \left(\frac{B}{B_0}\right)^2. \tag{3.8}$$

In turn, the normalized parameter B_0 depends on temperature as $(B_0)^2 \sim T^m$, where $m = 3p$ for 3d conductivity, p is the power in the temperature dependence of VRH resistance, Eq. (3.1). In our case, $p = 1/2$ and therefore $m = 3/2$:

$$B_0^2 = \frac{C_3 \eta^2}{e^2 a^4} \left(\frac{T}{T_{1/2}}\right)^{3/2}. \tag{3.9}$$

The numerical factor was estimated to be $C_3 = 288$ [6].

Let us discuss now the experimental results.

Small technical insert. The measurements were made on the series of Ge:As samples with donor impurities introduced by neutron-transmutation-doping (NTD), (2.28). To obtain n-type samples with small compensation, the Ge crystal was preliminary enriched up to 93.9 % with isotope ^{74}Ge. As a result, a series of Ge:As samples was obtained with As concentrations close to the metal–insulator transition and a small degree of compensation $K \approx 9$ %.

Figure 3.1 shows temperature dependence of resistance for one of measured samples with $N_{As} = 3.2 \times 10^{17}$ cm^{-3} in zero field and in fields $B = 1$, 3, 5 and 7 Tesla. One can see a large positive MR. At first sight, all the data are in reasonable agreement with the ES law, $p = 1/2$. In this approach, the magnetic field increases $T_{1/2}$ while the prefactor R_0 remains approximately constant. However, further analysis shows that the exponent p shifts to lower values with increasing B. We show below that this is due to the fact that in magnetic fields, the resistance curves reflect the smooth transition of the DOS from the Coulomb gap regime ($T^{-1/2}$ law) to a $T^{-1/4}$ law. The experimental data shown in Figs. 3.2 and 3.3 are in agreement with theoretical predictions (3.8) and (3.9): indeed, at low temperatures, $\ln[R(B)/R(0)] \sim B^2$ and $(B_0)^2 \sim T^{3/2}$.

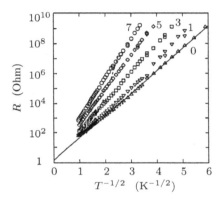

Fig. 3.1 The resistance of a sample of n-Ge:As on a logarithmic scale as a function of $T^{-1/2}$ in zero field and different magnetic fields [1]. Numbers on the curves indicate the magnetic field in Tesla.

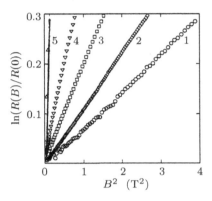

Fig. 3.2 Magnetoresistance in weak magnetic fields for the same sample as in Fig. 3.1 at different temperatures [1]: T, K: 1 — 1.04, 2 – 0.72, 3 – 0.50, 4 – 0.32, 5 – 0.10.

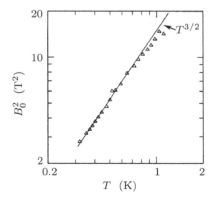

Fig. 3.3 Dependence of the parameter B_0 on T. The solid straight line shows the $T^{3/2}$ dependence.

Knowledge of $T_{1/2}$ and the slope of the straight line in Fig. 3.3 allows one to determine the value of a from Eq. (3.9).

In turn, it gives the values of κ and, finally, g_0. For the measured sample, we get $a = 22$ nm, $\kappa = 176$ and $g_0 = 1.2 \times 10^{15}$ cm^{-3} K^{-3}.

Another parameter which must be known is the half-width of the Coulomb gap Δ. The procedure for determining Δ involves plotting the experimental data $R(T)$ in the form of $I(E_c)$ vs. E_c, followed by a comparison with the theoretical curve (3.5) with the value of Δ as the only fitting parameter.

As mentioned above, the mean hopping distance $r(T) = [I(E_c)]^{-1/3}$. On the other hand, $r(T) = a\xi_c/2$ (see Ref. [5]), and the average hopping volume around each site is

$$V(\xi_c) = \frac{4\pi}{3} \left(\frac{a\xi_c}{2} \right)^3. \tag{3.10}$$

The number of sites which are critical for the formation of the percolation path is given by

$$m_c = I(E_c)V(\xi_c), \tag{3.11}$$

where $m_c = 7.66$ from percolation theory [5]. This gives

$$I(E_c) = \frac{6m_c}{\pi} \frac{1}{a^3} \frac{1}{(\xi_c(T))^3}. \tag{3.12}$$

In Eq. (3.12), the values m_c and a are known, the value of $\xi_c(T) = \ln[R(T)/R_0]$ can be determined at any temperature. The energy scale is given by $E_c = T\xi_c(T)$ and hence we can plot $I(E_c)$ vs. E_c near the Fermi level. Finally, we can then compare the experimentally derived $I(E_c)$ with a theoretical expression such as Eq. (3.5) to obtain the adjustable parameter Δ.

Figure 3.4 shows the results of such an analysis for the measured Ge:As sample. The chain curve is the cubic dependence (3.6) with the value g_0 determined in advance as described above, and the solid curve is the general dependence (3.5) with the fitting parameter $\Delta = 2.6$ K. The dashed curve is the asymptotic dependence (3.7), whose two parameters were determined in this way. It is also clear from Fig. 3.4 that the same curve fits the experimental data obtained in zero field and in all magnetic fields used in the experiment.

Note, that in strong magnetic fields, the shape of the electron wave function changes from spherical to a double paraboloid. This gives rise to correction coefficients in plotting the experimental dependence $I(E_c)$ which are described in the next section. One can see that only at lowest

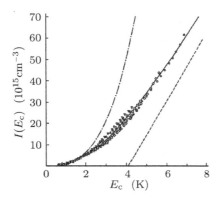

Fig. 3.4 Dependence of the integrated density of localized states $I(E_c)$ plotted vs. E_c [2]. The solid curve is (3.5) with $\Delta = 2.6$ K, the chain curve is (3.6), and dashed line is (3.7). The symbols correspond to magnetic fields shown in Fig. 3.1.

temperatures, $I(E_c)$ does exhibit a cubic dependence, which corresponds to the ES law. Increasing the temperature or the magnetic field shifts the experimental points away from cubic dependence, i.e., the experimental points lie in the crossover region. Also, one can see from Fig. 3.1 that even in strong fields, the curves $R(T)$ are still satisfactorily described by straight lines when plotted as $\ln R$ vs. $T^{-1/2}$. This seeming agreement is due to the limited temperature range and may lead to an erroneous interpretation.

Knowledge of g_0 and Δ allows one to determine $g(E_F) = g_0\Delta^2 \approx 8 \times 10^{15}$ cm^{-3} K^{-1} and to plot the entire $g(\varepsilon)$ curve for the DOS. This spectrum is shown in Fig. 3.5.

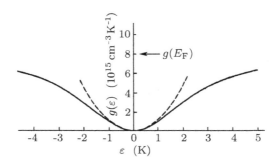

Fig. 3.5 The DOS $g(\varepsilon)$ for the measured sample determined by the "hopping spectroscopy" method [2]. The dashed line shows the "pure" Coulomb gap $g(\varepsilon) = g_0\varepsilon^2$.

The question arises of whether the obtained DOS spectrum is reliable in view of the fact that all the parameters were calculated on the basis of the selected model of the extended crossover, Eq. (3.4), when the values of Δ and, consequently, $g(\varepsilon)$ and $g(E_F)$ were obtained from Eq. (3.5). If one uses a different function for $g(\varepsilon)$ instead of (3.4), can one expect that the best agreement between the experimental and calculated functions will be obtained at a different value of Δ, which would then lead to a different value of $g(E_F)$? Our method is supported by an independent estimate of the value of $g(E_F)$.

Let us assume that all the impurity states N are randomly distributed with equal probability in an impurity band of width W, so $g(E_F) \approx N/W$. For the measured sample, $N = 3.2 \times 10^{17}$ cm^{-3}, which is close to the critical point of the metal–insulator transition. Near the transition, W is of order ε_1, the energy of ionization of electrons from impurity centers to the conduction band. According to Ref. [7], in Ge the value of ε_1 near the transition is about 4×10^{-3} eV, which gives $g(E_F) = 8 \times 10^{19}$ cm^{-3} eV^{-1} or 6.9×10^{15} cm^{-3} K^{-1}. This value is close to the value obtained on the basis of Eq. (3.4), especially taking into account that the DOS looks like a "Gaussian bell" rather than a rectangular box. It means that the smooth function $g(\varepsilon)$, Eq. (3.4), describes the real DOS accurately. We shall see below that in the case of sharper crossover from Mott to ES VRH, as is observed in the case of two-dimensional hopping conductivity, another function $g(\varepsilon)$ will be chosen for calculation of the DOS.

3.2 Two-dimensional Variable-Range-Hopping

In 2d, most experimental studies in the VRH regime have reported either Mott or ES behavior, (see, for example, Refs. [8, 9]). It should be noted that the temperature interval to observe ES VRH in gated 2d samples is limited at both low and high temperatures. The low-temperature limit of the ES law, predicted by Aleiner, Polyakov and Shklovskii [10], is connected with the screening of the Coulomb interaction by a metallic gate when the mean hopping distance $r(T)$ becomes larger than double the distance d to the metallic gate (in the VRH regime, the value of $r(T)$ increases continuously with decreasing T). As a result, at very low energies, the DOS again becomes constant: $g = \alpha(e^2/(\kappa d))$, where α is a numerical coefficient [10]. This low-temperature crossover phenomenon has been studied experimentally by Van Keuls et al. [11].

At high temperatures, a crossover from the ES regime ($T^{-1/2}$-law) to the Mott regime ($T^{-1/3}$-law) is caused by saturation of the DOS outside the Coulomb gap (CG) to its unperturbed value $g(E_F)$, which is observed with increasing T, when the width of the "optimal band" for VRH becomes larger than the width of the CG Δ. A further increase in T leads to smearing of the CG which we will describe in the next section. The high-temperature crossover phenomenon from ES to Mott VRH is universal and was first observed in 3d (see Chapter 2). It has been shown [12] that this crossover also occurs in 2d, in a δ-doped GaAs/Al$_{1-x}$Ga$_x$As heterostructure, when either the temperature or the carrier density are increased. We present a detailed study below of this crossover phenomenon which allows one to determine the DOS around the Fermi level and the width of CG Δ. All technical details concerning the sample structure and resistance measurements are presented in original paper [8].

The temperature dependence of the sample resistivity at a carrier density of 9.52×10^{10} cm^{-2} is plotted in Fig. 3.6 *vs.* $T^{-1/2}$ and $T^{-1/3}$ to check both laws for VRH conductivity.

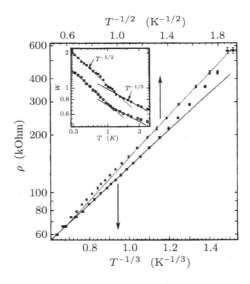

Fig. 3.6   The temperature dependence of the logarithm of the resistivity of δ-doped GaAs/Al$_{1-x}$Ga$_x$As sample with $n = 9.52 \times 10^{10}$ cm^{-2} in both $T^{-1/3}$ and $T^{-1/2}$ scales [9]. The inset shows the dimensionless energy of activation w *vs.* T on a log–log scale for samples with $n = 8.75$ (top) and 9.52×10^{10} cm^{-2}.

The low-temperature part of the curves is straight on the $T^{-1/2}$-scale, while the high-temperature part is straight on the $T^{-1/3}$-scale. In order to show the crossover more clearly, we plot in the inset the dimensionless energy of activation $w = -\partial[\ln \rho(T)]/\partial(\ln T) = p(T_p/T)^p$ as a function of T

on a log–log scale. The slope of the straight line on this scale is simply the exponent p, introduced in Eq. (3.1). Around 1.2 K, there is a sharp change of p from $1/2$ to $1/3$. In 2d, the crossover is sharp in contrast with 3d, where the crossover is smooth and holds over a wide range of temperatures (see Fig. 2.11 in Chapter 2). Correspondingly, the general form of $g(\varepsilon)$ has to be a sharp function with two limits: for a constant DOS at the Fermi level (FL), the integral density of states $I(E_c)$ must be a linear function of E_c:

$$I(E_c) = 2g(E_F)E_c, \tag{3.13a}$$

which is similar to the case of 3d, Eq. (3.3a). However, the second limit has to be different: in 2d, the CG is a linear function of energy: $g(\varepsilon) = g^*|\varepsilon|$, which leads to a quadratic dependence

$$I(E_c) = g^* E_c^2 \tag{3.13b}$$

in contrast to the cubic dependence for 3d conductivity, Eq. (3.3b). The analytical form of $g(\varepsilon)$ which satisfies both limits and corresponds to a sharp crossover can be written as

$$g(\varepsilon) = g^* \Delta \tanh\left(\frac{\varepsilon}{\Delta}\right). \tag{3.14}$$

Substituting Eq. (3.14) into Eq. (3.2) gives the expression for theoretical integral density of states $I^{\text{th}}(E_c)$:

$$I^{\text{th}}(E_c) = g^* \Delta^2 \ln[\cos(E_c/\Delta)], \tag{3.15}$$

which gives in the two limits

$$I^{\text{th}}(E_c) = g^* E_c^2, \qquad E_c \ll \Delta; \tag{3.16}$$

$$I^{\text{th}}(E_c) = 2g(E_F)(E_c - \Delta \ln 2), \qquad E_c \ll \Delta. \tag{3.17}$$

Equation (3.16) is equal to Eq. (3.13b) and Eq. (3.17) is equal to Eq. (3.13a) minus a term about $g(E_F)\Delta$ which arises because of a CG at the FL. The general expression (3.15) corresponds to the crossover phenomenon: a transition from a quadratic to linear dependence of $I(E_c)$.

Experimental values of $I(E_c)$ were obtained in a way similar to the case of 3d VRH with taking into account the features of two-dimensional conductivity. In 2d, the mean hopping distance $r(T) = [I(E_c)]^{-1/2}$. On the other hand, $r(T) = a\xi_c(T)/2$, where $\xi_c = (T_0/T)^p$ is the critical percolation parameter [4], and the average hopping area around each site is $S(\xi_c) = \pi(a\xi_c/2)^2$. The number of sites within this area, which is the critical for the formation of a percolation path, is given by $\chi_c = I(E_c)S(\xi_c)$, where $\chi_c = 3.2 \div 4.5$ for the different percolation models in 2d [5]. Therefore

$$I^{\exp}(E_c) = \frac{\chi_c}{S(\xi_c)} = \frac{4\chi_c}{\pi}\,(a\xi_c)^{-2}. \tag{3.18}$$

The value of $\xi_c = \ln[\rho(T)/\rho_0]$ can be obtained from the experimental data. The value of a was obtained from the slope $T_{1/2} = C_{\mathrm{ES}}(e^2/(\kappa a))$, taking $\kappa = 13.1$ as dielectric constant of GaAs (κ is not a function of a carrier density in 2d [11]). The latter is due to the peculiarities of the electric field distribution in a two-dimensional conducting plane, in contrast to the 3d conductivity, where the dielectric constant diverges as one approaches the critical concentration of the MIT. Inserting the expression for $T_{1/2}$ into Eq. (3.18) gives

$$I^{\exp}(E_c) = \frac{4\chi_c}{\pi C_{\mathrm{ES}}^2}\left(\frac{T_{1/2}}{(e^2/\kappa)\xi_c}\right)^2 = \frac{2g^*\chi_c}{C_{\mathrm{ES}}^2}\left(\frac{T_{1/2}}{\xi_c}\right)^2. \tag{3.19}$$

The numerical coefficient C_{ES} was determined by comparing of the theoretical value of $I^{\mathrm{th}}(E_c)$, Eq. (3.15), with the experimental values (3.19) at temperatures below 1 K. In this temperature interval, the ES law ($T^{-1/2}$-law) certainly holds (see inset in Fig. 3.6) and therefore $\xi_c^{\exp} = \xi_c^{\mathrm{th}}$. This gives $(\chi_c/C_{\mathrm{ES}}^2) = 0.5$, which corresponds to $C_{\mathrm{ES}} = 2.5 \div 3.0$ for $\chi_c = 3.2 \div 4.5$.

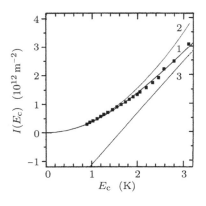

Fig. 3.7 The integrated density-of-states $I(E_c)$ *vs.* E_c [8]. Curves 1-3 correspond to Eqs. (3.15)-(3.17). The points are calculated from experimental data shown in Fig. 3.6.

Figure 3.7 shows the integrated DOS, $I(E_c)$, plotted as a function of E_c for the sample with $n = 9.52 \times 10^{10}$ cm^{-2}. The symbols are the experimental points calculated from Eq. (3.19), and were fitted to the theoretical expression Eq. (3.15) (solid line 1) to determine the parameter Δ. Curves 2 and 3 correspond to the limits $E_c \ll \Delta$, Eq. (3.16) and $E_c \gg \Delta$,

Eq. (3.17). The best fit yields $\Delta = 2.3 \pm 0.3$ K. The analysis of the curve at $n = 8.75 \times 10^{10}$ cm^{-2} gives the same value of Δ within the experimental uncertainty.

The crossover temperature T^*, which corresponds approximately to the point of the transition, can be estimated from the equality $\Delta = E_c(T^*) = T^*\xi_c$. It follows from Fig. 3.6 that crossover occurs at the resistivity of approximately 150 kOhm/sq, which corresponds to the value $\xi_c = 1.8$. This gives $T^* = \Delta/1.8 = 1.2$ K in agreement with experiment. It may be appropriate to mention here that if one uses the smooth function (3.4) for $g(\varepsilon)$ to analyze our crossover data for 2d VRH, one obtains $\Delta = 9.5 \pm 0.5$ K, which does not agree with experiment.

Using the values of $g^* = 0.4 \times 10^{12} m^{-2}$ K^{-1} (for $\kappa = 13.1$) and $\Delta = 2.3$ K, we calculated the DOS around the FL, Eq. (3.14), shown in Fig. 3.8. We plot the linear DOS only down to the temperature $T = 0.28$ K, as CG may be screened by the metallic gate at lower T.

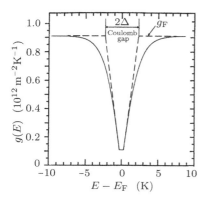

Fig. 3.8 DOS around the Fermi level for the sample with $n = 9.52 \times 10^{10}$ cm^{-2} [8]. The solid line corresponds to the function (3.14).

One thus sees that measurements of VRH resistance and MR (for 3d) can be used to determine the DOS spectrum around the Fermi level on the insulating side of the MIT. By analogy with the method of tunnel spectroscopy used for the same purpose on the metallic side of the transition, the method described above can be called "hopping spectroscopy". The key feature of this method is the precise identification of the low-temperature regime of the temperature dependence of the resistance $R(T)$ with the ES law, since in the crossover regime, the power p remains close to $1/2$ which may be a source of error.

3.3   Temperature-Induced Smearing of the Coulomb Gap

In what we discusses earlier, it is assumed that the DOS is constant; it does not depend on the temperature. The change of mechanisms of VRH is determined only by the distribution of electrons along the localized states due to energy transferred to the electron system. However, the DOS itself can be transformed because of smearing of the Coulomb gap with increasing temperature caused by screening of the Coulomb interaction with an increase of conductivity. This means that the deviation $R(T)$ from a straight line on the ES scale $\ln R$ *vs.* $T^{-1/2}$ with increasing T could be caused not only by a pure crossover from the ES to the Mott law, but also by the temperature-induced smearing of the Coulomb gap (CG).

The weakening of the gap has been predicted due to the filling of the localized states which results in more effective screening of the Coulomb interaction. This was demonstrated by computer simulation on the lattice model [13], on the random model of the impurity band [14], as well as by analytical calculations based on the self-consistent equation for the CG [15].

"Hopping spectroscopy" method allows one to detect this phenomenon experimentally. The results are described in Ref. [16]. The resistivity of two samples of ion-implanted Si:As with impurity concentration approximately 10 % below the metal–insulator transition has been measured for temperatures down to 0.1 K and in magnetic fields up to 11 T.

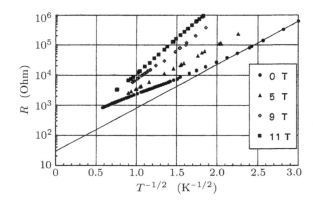

Fig. 3.9   The resistance of an ion-implanted Si:As sample in the VRH regime in zero field and in magnetic fields of 5, 9 and 11 T [16]. The solid line shows the fit to the $T^{-1/2}$ law in the low-temperature limit.

Figure 3.9 shows the results of the resistance measurements as a function of temperature on the scale $\ln R$ *vs.* $T^{-1/2}$ in different magnetic fields. The key point of our analysis is the assumption that the ES law in the VRH conductivity is only reached at lowest temperatures below 200 mK. From the slope of the straight line, we obtain $T_{1/2} \equiv T_{ES} = 11.2$ K, and the intercept gives $R_0 = 26.7$ Ohm. The deviation of the resistance above the straight line with increasing T is interpreted as the crossover regime. Since $T_{1/4} \equiv T_M = 420$ K for this sample [17], one can obtain from Eqs. (2.30)–(2.34) the value of the half-width of the CG $\Delta = 1.8$ K:

$$\Delta = \sqrt{\frac{g(E_F)}{g_0}} = \sqrt{\frac{T_{ES}^3}{T_M}}. \tag{3.20}$$

Let's determine the DOS from the experimental data. In strong magnetic fields, Eq. (3.10) has to be modified because of the shrinking of the electron wave function and the change of its shape from a sphere to a double paraboloid. This leads to an additional factor $X(s)$ in expression for $V(\xi_c)$, cf. Eq. (3.10):

$$V(\xi_c) = \frac{4\pi}{3} \left[\frac{a\xi_c(B,T)}{2} \right]^3 X(s), \tag{3.21}$$

where $\xi_c(B,T) = \ln[R(B,T)/R_0]$ and $s = (\pi e/h)a^2 B\xi_c(B,T)$. Function $X(c)$ is shown in Fig. 3.10.

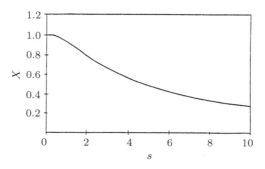

Fig. 3.10 The function $X(s)$ [1].

Consequently,

$$I(E_c) = \frac{6m_c}{\pi a^3} \frac{1}{[\xi_c(B,T)]^3 X(s)}. \tag{3.22}$$

One can calculate the normalized integrated DOS $I^*(x) = I(x)[(2/3)g_0\Delta^3]^{-1}$, where $x = E_c/\Delta$ is the dimensionless energy. The value

$a = 16$ nm is used in calculation, which is quite reasonable for the given impurity concentration in Si:As [18]; this is the only adjustable parameter. The quantities $\xi_c(T)$ and $\xi_c(B,T)$ were both determined from experimental data, assuming that R_0 is constant and independent of magnetic field. The value of $\Delta = 1.8$ K.

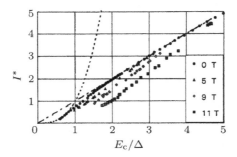

Fig. 3.11   The normalized integrated DOS $I^*(x)$ *vs.* dimensionless energy $x = E_c/\Delta$ [16].

The results are shown in Fig. 3.11. The dotted curve corresponds to the cubic dependence $I^*(x) = x^3$ for the pure parabolic gap, while the dot-dashed line shows the linear dependence $I^*(x) = x$ for the constant DOS without the Coulomb gap, Eq. (3.3a). This is in contrast to the data for the sample of Ge:As shown in Fig. 3.4, where the asymptotic linear dependence follows Eq. (3.7) with a negative term $\pi g_0 \Delta^3$ due to the presence of the CG at the Fermi energy. This means that sizable deviations of the resistance above straight line with increasing T in zero magnetic field (Fig. 3.6) are caused by the temperature-induced smearing of the CG rather than from a crossover from the ES to the Mott law. It can hardly be explained in the framework of the "pure" theory because the temperature changes in DOS are in the temperature interval αT, where $\alpha = $ const., while the half-width of the optimal band is $E_c = (T_{ES}T)^{1/2}$. Therefore, $E_c/T = \xi_c = (T_{ES}/T)^{1/2} \gg 1$ and so it would seem that the temperature increase cannot influence VRH conductivity by changing the DOS. However, because α is relatively large ($\alpha = 3$) [19], of order ξ_c, it is possible to observe the effect of temperature-induced smearing.

We note that information on the DOS was obtained indirectly via the integrated DOS. Therefore the smearing of the CG is given only qualitatively. However, it is the first experimental verification of this effect.

One can also argue that the integrated DOS is essentially the concentration of the localized states. If so, how can it depend on temperature? Indeed, the DOS integrated over the entire energy range is just the total concentration of localized states in the impurity band. However, only those states with energy level inside the optimal band of the width $2E_c$ contribute to the VRH conductivity. Therefore, an important question is how the states are redistributed between different energy ranges with increasing T. The above method of "hopping spectroscopy" of the integrated DOS cannot answer this question. In order to study this problem, a Monte Carlo computer simulation of the DOS in the presence of the Coulomb gap at $T = 0$ was performed for different T. The algorithm of the simulation is described in details in Ref. [20].

The results for the DOS at intermediate degree of compensation $K = 0.5$ are shown on Fig. 3.12(a) for different temperatures. The values of T, as well as other energies, are given in units of $(e^2/\kappa)N^{1/3}$, which is the Coulomb energy on the mean inter-impurity distance (N is the concentration of the majority impurities and κ is the dielectric constant of the semiconductor). The energies are plotted with respect to the Fermi energy E_F, $\varepsilon = E - E_F$. A pronounced CG is seen at $T = 0$, which smoothly disappears with increasing T. The hard gap in the DOS at $T = 0$ is due to the relatively small size of the simulated array (100 donors). Being interested only in the evolution of the DOS with increasing T, the results of the computer simulation were not averaged over different arrays to obtain a smooth curve. At non-zero T, such averaging is provided by the temperature-induced hopping movements of electrons between the sites.

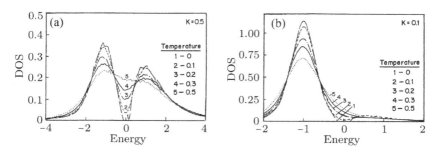

Fig. 3.12 The computer simulation of the temperature-induced smearing of the Coulomb gap for lightly doped semiconductors with compensation $K = 0.5$ (a) and $K = 0.1$ (b) [16]. The temperature and energies are shown in the units of $(e^2/\kappa)N^{1/3}$.

The results of simulations of the system having low compensation $K = 0.1$ are shown in Fig. 3.12(b). The main feature of these results is the significant change of the DOS not only in the region of the CG, but also near the maximum of the distribution in spite of relatively low T, much less than the energy distance between the maximum of the distribution and the position of the Fermi level. This means that even at low temperatures, the redistribution of states is so effective that some additional states come into the region of the optimal band E_c with increasing T. This permits the increase of $I(E_c)$ to be observed experimentally. The result holds until the width of the optimal band E_c remains small compared with entire width of the impurity band ($E_{band} \approx 1$ in the units of Fig. 3.12). Moreover, if E_c becomes comparable with E_{band}, the VRH regime converges into the regime of nearest-neighbor hopping (NNH).

3.4   Hopping   Conductivity   in   Semiconductor   Solid Solutions

To conclude this chapter, we will show how the measurements of hopping conductivity allow one to clarify the mechanism of the influence of the fluctuations in solid solution composition on the value of scatter of impurity levels.

The energy of activation of NNH conductivity, ε_3, is influenced strongly by the energy scatter of impurity levels. In the case of lightly doped semiconductors, the scatter is due to random electrostatic fields of charged impurities. In the case of a solid solution, there may be an additional scatter due to fluctuations of the compositions. Measurements of the hopping conductivity may provide a very sensitive method for investigating these fluctuations. The experimental results presented below show that even when the concentration of silicon in germanium is several percent, the activation energy of hopping conductivity is governed entirely by fluctuations of the solid solution composition. To study this problem, hopping conductivity was measured on doped samples of $Ge_{1-x}Si_x$ solid solutions with $0 < x < 0.08$. The results were presented in Ref. [21].

Small technical insert. The selection of $Ge_{1-x}Si_x$ solid solutions was not accidental. In order to study the influence of the composition on the broadening of the energy interval of impurity levels, it was necessary to dope all the samples with approximately the same amount of shallow impurities, and it was desirable to ensure a constant degree of compensation close to $K \approx 0.5$ at which ε_3 should have its minimal value [5]. This was achieved by means of the neutron-transmutation-doping method (NTD, see Eq. (2.28) in Chapter 2), irradiating a batch of $Ge_{1-x}Si_x$ samples with slow neutrons in a nuclear reactor. Irradiation of germanium leads to appearance of gallium as acceptors and arsenic and selenium as donors; in this way, p-type samples with $K = 0.4$ are obtained. The concentration of the majority impurity, gallium, is calculated from the expression $N_{Ga} = \sigma\Phi$, where Φ is the total thermal neutron dose (cm^{-2}) and $\sigma = 3.2 \times 10^{-2}$ cm^{-1} is the reaction coefficient [22]. Irradiation of silicon produces phosphorous donors, but the coefficient of this reaction is two orders of magnitude lower, so that the concentration of phosphorous in samples with x less than 0.06 could be ignored. All samples, including a control pure Ge crystal, received an approximately equal neutron dose $\approx 10^{17}$ cm^{-2}, which corresponds to a concentration of Ga acceptor impurity in the narrow interval $N_{Ga} \approx 3.5 \times 10^{15}$ cm^{-3} and the same degree of compensation $K \approx 0.40$.

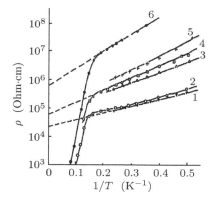

Fig. 3.13 Temperature dependence of the resistivity of NTD p-$Ge_{1-x}Si_x$ samples [21]. x: 1 – 0 (pure Ge), 2 – 0.7, 3 – 2.2, 4 – 3.4 ÷ 3.7, 5 – 4.15, 6 – 6.0 ÷ 6.3.

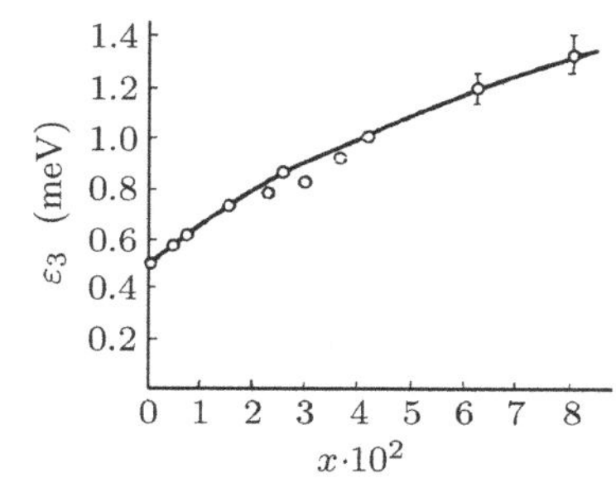

Fig. 3.14 Dependence of the hopping energy of activation $\varepsilon_3(x)$ on the solid solution composition [24]. The solid line is the result of calculation (3.24).

Figure 3.13 shows the temperature dependences of the resistivity of the investigated $Ge_{1-x}Si_x$ samples. The hopping energy of activation of conductivity ε_3 increases considerably with increasing silicon fraction x. This dependence is shown in Fig. 3.14. The value of $\varepsilon_3 = 0.5$ meV for pure Ge ($x = 0$) is in good agreement with the published values for p-Ge:Ga with $K \approx 0.4$ (see Fig. 2.8 in Chapter 2). One can see also from Fig. 2.8 in Chapter 2, that ε_3 depends weakly on N_{Ga} and therefore all changes of ε_3 in our $Ge_{1-x}Si_x$ samples are determined by the increase of x.

Scatter of the impurity levels in solid solutions is determined by fluctuations in the position of the band edges (for p-Ge, the energy levels of shallow Ga acceptors are closely connected with the position of the edge of the valence band).

Composition fluctuations of all possible scales may occur in a solid solution. The best theoretical description can be provided for large-scale fluctuations extending over volumes containing many atoms of all the components of the solution. In this case, the fluctuations can be described by a composition parameters $x(r)$ which varies smoothly in space. The band parameters at the point r are equal to the average values for a solid solution with $x = x(r)$. This means that if the average value (over some volume) of the concentration differs by δx from the average concentration over the whole crystal x^*, the valence (or conduction) band edge changes in this volume by an amount $\delta E_V = \alpha \delta x$, where $\alpha = dE_V/dx|_{x=x^*}$. (The value of α can be estimated if we know the variation of the valence band during the transition from pure Ge to pure Si. The value $\alpha = 0.21$ eV was obtained from investigation of the Ge–Si heterostructure [23].)

The smaller the volume under consideration, the greater the fluctuations of the composition. It is also obvious that an acceptor level is not sensitive to fluctuations whose size is less than the acceptor localization radius a_B. Therefore, the main contribution to the shift of the level comes from fluctuations of the size of order a_B. The average number of Si atoms in such a volume is Nxa_B^3, where N is the density of atoms in the Ge lattice and a typical fluctuation of this magnitude is of order $(Nxa_B^3)^{1/2}$. Therefore, a typical fluctuation of x is $\delta x = (Nxa_B^3)^{1/2}/(Na_B^3)$ and shift of an acceptor level is approximately equal to

$$\varepsilon_3(x) = \alpha\sqrt{\frac{x}{Na_B^3}}.$$

The detailed theoretical calculation of $\varepsilon_3(x)$ presented in Ref. [24] gives the following expression:

$$\varepsilon_3(x) = \frac{0.54}{2\sqrt{\pi}}\frac{\alpha\sqrt{xN}}{N\sqrt{a_B^3}} = 0.15\frac{\alpha\sqrt{x}}{\sqrt{Na_B^3}} = A\sqrt{x}. \tag{3.23}$$

Substituting $a_B = 4$ nm, $\alpha = 0.21$ eV, $N = 4.5 \times 10^{22}\,\text{cm}^{-3}$ into Eq. (3.23), we find that $\varepsilon_3(x) = 0.18$ meV for $x = 0.06$, which is several times smaller than the experimentally observed values (Fig. 3.14).

The solution of this discrepancy was suggested in Ref. [24]. In the previous consideration, it was assumed that Si atoms are distributed in Ge lattice randomly, separately and there is no correlation in their position. However, fluctuations will be much stronger if Si atoms penetrate into Ge lattice not separately, but form clusters of K atoms in average. If there is no correlation in the space distribution of clusters, the previous theoretical calculations are valid with the only replace x by Kx.

The experiment shows (Fig. 3.14) that in our interval of x, the value $\varepsilon_3(x)$ caused by the composition fluctuation, is of the same order as $\varepsilon_3(0)$.

The latter is due to the Coulomb scatter of impurity levels in compensated Ge.

When analyzing the data, must bear in mind that $\varepsilon_3(0)$ in any case should not be considered as an addition to Eq. (3.23). If these two kinds of fluctuation are independent, one can wright the following expression:

$$\varepsilon_3^{\exp} = \sqrt{\left[\varepsilon_3^{\text{theor}}\right]^2 + \left[\varepsilon_3(0)\right]^2}. \tag{3.24}$$

Figure 3.14 shows the result of calculation of $\varepsilon_3(x)$ using Eq. (3.24), $\varepsilon_3(0) = 0.5$ meV and $\varepsilon_3^{\text{theor}} = Ax^{1/2}$. The best agreement corresponds to $A = 4.2$ meV. Using $A = 0.15\alpha K^{1/2}/(Na_{\text{B}}^3)^{1/2}$, one obtain $K \approx 50$. Of course, the value of $\alpha = \mathrm{d}E_{\text{V}}/\mathrm{d}x$ at low x can slightly differ from its average value 0.21 eV [23], but this cannot change the main conclusion that $K \gg 1$ and Si atoms go into Ge as clusters of K atoms together. Control measurements on other alloys show that the value of K fluctuates from 30 to 150, which indicates its technological origin.

Chapter 4

Spin-dependent and "Phononless" Hopping Conductivity

In the theoretical analysis of hopping conductivity, the spin–spin exchange interaction between localized electrons was neglected, in contrast to the Coulomb interaction. This is reasonable for a small concentration of impurities because the exchange interaction drops exponentially with increasing distance while the Coulomb interaction decreases more slowly, as a power-law. However, in the case of heavily doped and compensated (HDC) semiconductors, the distance between donors became smaller which makes it possible to observe phenomena associated with the spin exchange interaction. These effects are discussed below.

4.1 "Magnetic" Hard Gap at the Fermi Level

Let us recall the sequence of the hopping mechanisms with decrease of temperature. In general, the temperature dependence of hopping resistivity has the form:

$$\rho(T) = \rho_0 \exp\left(\frac{T_p}{T}\right)^p. \tag{4.1}$$

At relatively high temperatures, a nearest-neighbor hopping conductivity (NNH) dominates with constant energy of activation ($p = 1$). With decreasing T, NNH is replaced by the variable-range hopping (VRH) among localized states having energies close to the Fermi level (FL). The Coulomb interaction between the localized carriers leads to a soft Coulomb gap (CG) in the density-of-states (DOS) around the FL. When T exceeds the half-width of the CG Δ ($T > \Delta$), $\rho(T)$ obeys the "Mott" law ($p = 1/4$ for 3d and $p = 1/3$ for 2d conductivity). When $T < \Delta$, $\rho(T)$ is described by the "Efros–Shklovskii" (ES) law with $p = 1/2$ (both for 3d and 2d).

Does the ES law represents the last conductivity mechanism as $T \to 0$? The answer depends on whether the CG retains its form (parabolic in 3d and linear in 2d) down to the lowest energy ε ($\varepsilon = 0$ at the FL). The point is that the Coulomb interaction implies a correlation between energies and distances of any two states above and below the Fermi level (see Eq. (2.21)):

$$\varepsilon_j - \varepsilon_i - \frac{e^2}{\kappa r_{ij}} > 0, \qquad (4.2)$$

where the state i below the Fermi level is initially occupied, the state j above the Fermi level is initially empty, and r_{ij} is the distance between these two states. It follows from (4.2) that occupied and empty states must be separated energetically by at least the Coulomb energy corresponding to their distance. In order to retain the ES law, it is necessary that the correlation described in (4.2), persists at long distance. However, at $T \to 0$, the hopping distance $r(T)$ increases monotonically and may eventually become larger than the screening radius. When this happens, the CG will be smeared which results in the reverse crossover from the ES law back to the Mott law with decreasing T. An example of such screening of the CG was discussed in Chapter 3 where it was mentioned that in 2d systems, for which the carrier concentration is controlled by metallic gate located on distance d from the two-dimensional conductive layer, the Coulomb interaction is screened at temperatures when the VRH hopping distance $r(T) > d$. Experimentally, the reverse transition from the ES law back to the Mott law with decreasing T manifests itself as a deviation from the straight line on a plot of $\lg \rho$ vs. $T^{-1/2}$ to *lower* resistance.

There is, however, another observation when $\rho(T)$ deviates from the $T^{-1/2}$ law to *higher* resistances with decreasing T and is replaced by a stronger temperature dependence with $\rho = 1$ [1–3]. Such a deviation was first observed in samples of amorphous germanium–chromium films [1] and in polymer films irradiated with high-energy ions [2]. The interpretation of this effect was based on the assumption of a hard "magnetic" gap at the Fermi level caused by the spin–spin exchange interaction energy J and the necessity for electrons to overcome the average spin exchange energy J at each hop. In this case, for $T < J$, the $T^{-1/2}$ dependence will be replaced by a stronger linear law, characteristic of the conductivity with constant energy of activation.

In the first experiments, a series of $Ge_{1-x}Cr_x$ amorphous films was specially prepared to study the influence on hopping conductivity the magnetic impurity-chromium, which has an uncompensated magnetic moment. The details of fabrication $Ge_{1-x}Cr_x$ amorphous films are described in Ref. [1]. Figure 4.1 shows $\sigma(T)$ for films of $Ge_{1-x}Cr_x$ for different x. At a high concentration of Cr atoms, $x > 0.12$, the conductivity is metallic-like and temperature-independent, while for $0.08 < x < 0.10$, $\rho(T)$ obeyed the VRH law, when the energy of activation $\varepsilon(T)$ continuously decreases with decreasing T. However, below 1 K, $\varepsilon(T)$ no longer decreases, which corresponds to a transition to conductivity with constant energy of activation. It was suggested that this transition could be explained by influence of a hard "magnetic gap" which was discussed above. One can verify this suggestion: in strong magnetic fields, the deviation from $T^{-1/2}$ law to T^{-1} law must disappear, because all electron spins are oriented along the field, so the energy change is equal for all states and thus should not be taken into account. Indeed, for the sample $x = 0.087$, in strong magnetic field, no deviation from the VRH law is observed (dashed line in Fig. 4.1) and $\varepsilon(T)$ continues to decrease with T down to the lowest measured temperature.

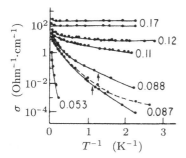

Fig. 4.1 Hopping conductivity of amorphous films $Ge_{1-x}Cr_x$ as a function of temperature [1]. Numbers on curves show the value of x.

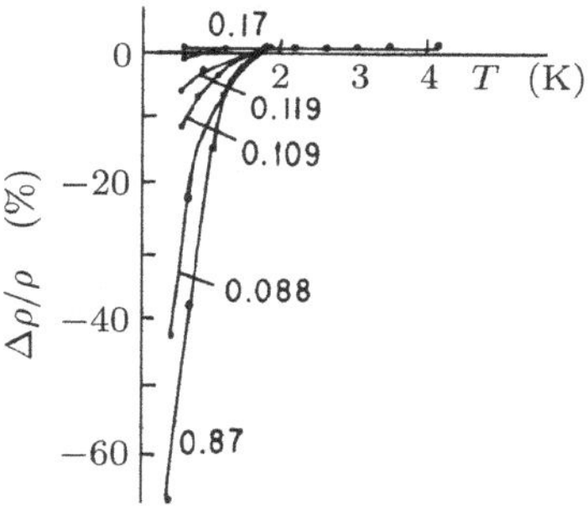

Fig. 4.2 Temperature dependence of the magnetoresistance of $Ge_{1-x}Cr_x$ films for different x at $H = 10$ kOe [1].

The resistance difference between stronger "T^{-1}" and weaker "$T^{-1/2}$"-dependences increases rapidly. Therefore, we could expect a considerable increase of the conductivity in magnetic field with decreasing T. Figure 4.2 shows the magnetoresistance (MR) $\Delta\rho/\rho$ of the $Ge_{1-x}Cr_x$ films with different x in magnetic field of 1 Tesla as a function of T. One can, indeed, see that at $T < 1$ K, the $\Delta\rho/\rho$ changes its sign from positive to negative and the magnitude of the effect increases rapidly with decreasing T.

A similar effect was observed on samples of polymer films in which irradiation with energetic ions led to metallic-like conductivity [2]. Resistivity

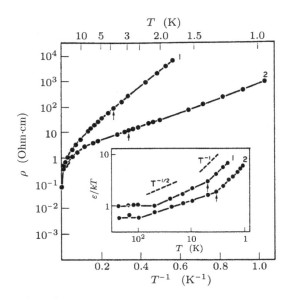

Fig. 4.3   Temperature dependence of resistivity of polymer films irradiated with different doses of energetic ions [2]. Φ, cm^{-2}: 1 — 3×10^{16}, 2 — 10^{17}. The insert shows the reduced energy of activation *vs.* temperature on log–log scale.

of two polymer films irradiated with Ar$^+$ ions are shown in Fig. 4.3. One sees that in the temperature range 100–4 K, the hopping energy of activation $\varepsilon(T)$ continuously decreases obeying the ES law with $p = 1/2$. For $T < 4$ K, the energy of activation no longer decreases which corresponds to a transition to conductivity with constant energy of activation $\varepsilon = 4 \div 0.2$ meV depending on the irradiation dose. In the same temperature interval, an anomalously large negative MR was observed (Fig. 4.4), similar to the case of Ge$_{1-x}$Cr$_x$ films.

In these samples, the deviation from $T^{-1/2}$ law to T^{-1} law was unexpected because no impurities with uncompensated magnetic moments were introduced. Probably, the spin–spin exchange interaction was provided by a specific space distribution of wave functions for the electrons localized on the polymer backbones. The relatively high temperature of the assumed magnetic ordering effect ($T \approx 4$ K) allowed one to the measurements of the electron spin susceptibility by means of electron spin resonance (ESR). The result is shown in Fig. 4.5. At relatively high temperatures, where $T^{-1/2}$ law is observed, the susceptibility is paramagnetic and follows the Curie law. At $T < 4$ K, a deviation from the Curie law is observed, which can be considered a direct evidence for magnetic ordering of ferromagnetic character. Moreover, our estimate of the value of the exchange integral agrees with the resistivity activation energy. Thus, the measurements of the magnetic susceptibility confirm the model of a "magnetic" hard gap in the DOS induced by the spin–spin interaction.

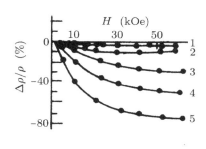

Fig. 4.4 Negative magnetoresistance of polymer sample 2 for different temperatures [2]. T, K: 1 — 2.1, 2 — 1.54, 3 — 0.97, 4 — 0.66, 5 — 0.47.

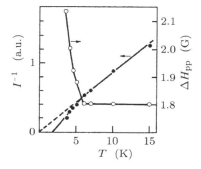

Fig. 4.5 Temperature dependence of the intensity and the width of the ESR signal for polymer sample 2 [2].

The deviations from $T^{-1/2}$ law to a constant energy of activation with decreasing T were also observed in other materials: in a dilute magnetic semiconductor $Cd_{0.91}Mn_{0.1}Te:In$ [4], in ion-implanted Si:As [5], in partially compensated ion-implanted Si:P, B [6], in n-^{74}Ga:As [3] and in p-Si:B [7]. Figure 4.6 shows the resistivity of a Si:B sample plotted as lg ρ *vs.* $T^{-1/2}$ (a) and *vs.* T^{-1} (b) in zero magnetic field [7]. The straight line in Fig. 4.6(a) demonstrates that the ES law is obeyed only in the middle interval of temperatures and there are two deviations from the $T^{-1/2}$ law: at high temperatures and at low temperatures.

Figure 4.6(b) shows that at lowest temperatures, the resistivity obeys a T^{-1} law. The deviation to this law disappears in a strong magnetic field. In view of this, one can conclude that in zero magnetic field, a hard gap appears at the bottom of the DOS which is destroyed in a strong magnetic field. This indicates that the gap is indeed of magnetic origin, as was suggested earlier in Ref. [3]. The magnetic field affects the hopping transport through its effect on the localized magnetic moment of the holes that are known to exist in p-Si:B [8].

In Fig. 4.6(a), the high-temperature deviation of the resistance above the straight line on $T^{-1/2}$-plot is determined by the crossover to the Mott

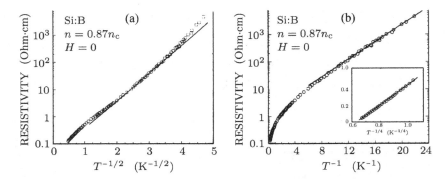

Fig. 4.6   The resistivity of insulating Si:B plotted on a log scale as a function of $T^{-1/2}$ (a) and T^{-1} (b). The inset to (b) shows data at higher temperatures plotted as a function of $T^{-1/4}$ (adopted from Ref. [7]).

law (see inset of Fig. 4.6(b) plotted as a function of $T^{-1/4}$). In this case, all transformations of the VRH conductivity with decreasing temperature are shown. At high temperatures, when T exceeds the half-width of the CG Δ, the Mott law is observed, whereas for $T < \Delta$, the Mott low is replaced by ES law ($T^{-1/2}$-law), and finally, when T is less than a magnetic "hard gap", $T < J$, the $T^{-1/2}$-law is changed to the T^{-1}-law.

There is no theoretical work that supports the suggestion that taking into account the spin interaction between localized electrons and the broad distribution of exchange energies will result in the appearance of a "magnetic" hard gap at the Fermi level. The role of spin–spin correlations in the VRH regime still needs further theoretical investigation.

4.2   Hopping Negative Magnetoresistance

There is another aspect of hopping conductivity which requires taking into account the spin interaction of localized electrons. We bear in mind the effect of negative magnetoresistance (NMR). Theoretical consideration of the influence of a magnetic field on hopping conductivity predict the increase of resistance (positive magnetoresistance, PMR) due to the shrinkage of the electron wave function in the plane perpendicular to the direction of magnetic field. All theoretical predictions about PMR were confirmed by experiments both in the NNH and the VRH regimes (see Chapter 3, Eqs. (3.8), (3.9) and Figs. 3.2 and 3.3, and also Ref. [9] and references therein). However, in heavily doped, compensated (HDC) Ge samples with the Mott VRH conductivity ($T^{-1/4}$-law), the opposite effect of NMR has been observed [10], with the resistivity slightly decreases in weak magnetic fields (Fig. 4.7). The effect of NMR was also observed on the metallic side of the metal–insulator transition. However in this case, there are theoretical explanations of NMR based on the suppression in magnetic fields of quantum corrections for diffusive conductivity of delocalized electrons [11] or spin-dependent electron scattering on magnetic impurities [12]. It seems that both ideas are not applicable to localized electrons with hopping mechanism of motion where each hop is inelastic, accompanied by the absorption or emission of a phonon which breaks the phase coherence. Therefore, it is necessary to consider new ideas to understand the effect of NMR in the hopping regime [13].

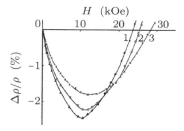

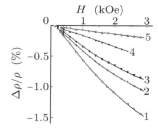

Fig. 4.7   Magnetoresistance of a HDC-Ge sample at different temperatures [10]. T, K: 1 — 3.0, 2 — 3.5, 3 — 4.2.

Fig. 4.8   Negative magnetoresistance of HDC-Ge sample as a function of temperature at weak magnetic fields [13]. T, K: 1 — 2.1, 2 — 3.5, 3 — 4.2, 4 — 7.2, 5 — 11.5.

Experiment shows that the characteristic feature of the NMR effect in the VRH regime is the absence of a quadratic dependence on magnetic field even in very weak fields H when $\mu^* H < T$ (here μ^* is the normalized magnetic moment of an electron, including the Lande-factor), Fig. 4.8. The explanation of NMR in Mott type VRH conductivity suggested by Efros [13] was based on two assumptions:

(i) there exists a ferromagnetic ordering of spins of localized electrons caused by their exchange interaction. The characteristic energy J of this exchange interaction (Curie temperature) must exceed both T, as well as the energy of the "optimal" band around the Fermi level, where VRH conductivity is implemented;

(ii) there are also states with coupled spins in the vicinity of the Fermi
   level (FL). For example these may be two states which are close to
   each other and separated from other states.

It can be assumed that for such "twins", a strong antiferromagnetic
interaction appears which leads to spin pairing. Due to the ferromagnetic
interaction in a very weak field, unpaired spins of "single" states are oriented
along the field. The corresponding energies decrease with increasing H.
However, if there are only "singles" near the FL, the DOS will be not
changed because the Fermi level itself decreases together with the electronic
states. Let us now consider the levels which appear after ionization of a
"twin". We note that the remaining spin must be oriented along H because
the energy level which corresponds to ionization of the "twin" with electron
spin oriented against the field, lies below on the energy interval equal to
the exchange interaction of "singles".

Thus, an electron from a "twin" that participates in hopping transport
has its spin oriented against field. In other words, the level of the "twin"
shifts up with increase of H. This results in a shift in the position of the
FL and to a corresponding change in the DOS $g = g(E_\mathrm{F})$. It follows from
general consideration, that

$$\frac{\Delta g}{g} = \alpha \frac{\mu H}{\varepsilon}. \tag{4.3}$$

The sign of the numerical coefficient α and the energy ε are determined by
the density of states for "singles" and "twins" and their energy derivatives.
In the case of the Mott VRH, where $\rho = \rho_0 \exp(T_0/T)^{1/4}$ and $T_0 \approx (ga^3)^{-1}$,
this leads to a change in T_0 and therefore, in resistivity:

$$\frac{\Delta \rho}{\rho} = -\frac{1}{4} \frac{\Delta g}{g} \ln\left(\frac{\rho}{\rho_0}\right). \tag{4.4}$$

Inserting(4.3) into (4.4) yields

$$\frac{\Delta \rho}{\rho} = -\frac{\alpha}{4} \frac{\mu H}{\varepsilon} \ln\left(\frac{\rho}{\rho_0}\right). \tag{4.5}$$

Equation (4.5) predicts that the value of NMR $\Delta\rho/\rho$ must be proportional to $\lg \rho$. Experimental results are in agreement with this prediction
(Fig. 4.9).

The mechanism of NMR described above is not valid for the ES VRH,
because in this case $g(E_\mathrm{F}) = 0$. We note that in this model, not only NMR

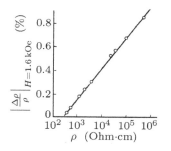

Fig. 4.9 The relationship between the values of NMR of HDC-Ge sample at $H = 1.6$ kOe and logarithm of sample resistivity at different temperatures [13].

but also PMR is possible. There is another consequence of the proposed model: saturation of the NMR effect in strong magnetic fields, when all "twins" will be ionized and therefore transformed into "singles". Experimentally, saturation is usually masked by positive MR (Fig. 4.7). However, saturation of NMR was revealed after subtracting PMR (proportional to B^2) from the total MR curves [14].

Recently, Agam, Aleiner, and Spivak [15] suggested a new "spin memory" mechanism for isotropic NMR in hopping regime in weak magnetic fields. The model is based on the assumption that the characteristic hopping time is much shorter than the spin relaxation time and NMR is due to spin memory effect when the hopping rate of an electron between sites i and j depends on the spin configuration of the hopping electron and a spin of electron located nearby at site ij which can provide for an indirect transition via a virtual state of double occupancy (Fig. 4.10). If both spins form a triplet state, the indirect transition is suppressed, while in magnetic field due to local fluctuations of the Lande-factor in strongly disordered system, the spin correlation is destroyed which may lead to NMR.

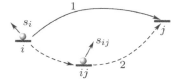

Fig. 4.10 A simplified hopping model in which electron may hop directly between two sites (1) or go indirectly (2) via nearby occupied site forming virtual singlet state of double occupancy [15].

We have already noted that each electron hop is accompanied by phonon absorption or emission. Therefore, the electron loses its phase after each hop, and there is no interference between successive hopes. However, more recent theoretical studies show that this statement is correct only for nearest neighbor hopping (NNH). In the VRH mechanism, the electron hops

over a long distance passing near centers which are occupied by electrons
(Fig. 4.11). It was shown by Shklovskii and Spivak [16] and Ioffe and Spivak
[17] that the probability of a long-distant hop is determined by the interference of many paths of the tunneling which include scattering processes.
Each impurity i participates in scattering which results in a spherical wave
with exponential decay

$$\Psi_{\text{scat}} = \psi_1^0(r_i)\frac{\mu_i}{4\pi|r - r_i|}\exp\left(-\frac{|r - r_i|}{a}\right), \tag{4.6}$$

where $\mu_i \equiv 8\pi a\varepsilon_i/(\varepsilon_1 - \varepsilon_i)$ is the scattering amplitude or, in other words,
the scattering length and $\varepsilon_i < 0$ is the energy of impurity state i. An
electron can be scattered many times on its way from 1 to 2. All these
waves, together with the non-scattered wave, contribute additively to the
amplitude of the wave function $\Psi_1(r_2)$ which reflects the probability for
electron localized on site 1 to appear on site 2. Because of the exponential
decay, only the shortest paths contribute to $\Psi_1(r_2)$. They are concentrated
inside the cigar-shaped domain (Fig. 4.12).

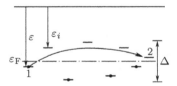

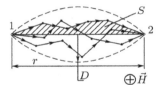

Fig. 4.11 Levels of impurities located
near the straight line connecting centers
1 and 2. The chain line shows the position of the Fermi level and straight line
indicates the bottom of the conduction
band. The electron hop from center 1
to center 2 is shown by the arrow [16].

Fig. 4.12 Three zigzag paths con-
tributing to the probability of a hop
from center 1 to center 2. The sigar-
shaped region with the transverse diam-
eter D containing all important paths is
shown by broken lines. The area S attributed to one path is shaded [16].

Along these paths, the electron is scattered only in the forward direction.
The amplitudes, μ_i, describing an individual scattering process at state i
may be positive or negative. The interference can be constructive (when all
scattering amplitudes μ_i are positive) or destructive (when some of them are
negative). A magnetic field applied perpendicular to the area S formed by
a zigzag path and the straight line between sites 1 and 2 (which corresponds
to the non-scattered transition) will destroy the interference. Therefore, the
sign distribution of μ_i determines the sign of MR, and the application of a

weak magnetic field to a macroscopic sample may lead to either decrease or to an increase of the conductivity.

This mechanism of NMR could be called "orbital MR", because it is not determined by the electron spin interaction. For two-dimensional conductivity, however, all hops are in the plane. As a result, the "orbital" mechanism of NMR may be observed only if there is a component of the applied magnetic field perpendicular to the plane. By contrast, the effect of MR determined by spin alignment is isotropic. Therefore, measurements of hopping MR in 2d samples in parallel magnetic fields give direct evidence of the influence of spin polarization on hopping conductivity.

Such studies have been conducted on a single 2d layer in a δ-doped GaAs/Al$_x$Ga$_{1-x}$As heterostructure with a VRH mechanism of conductivity. The results of this study are presented in Ref. [18], where details are given about sample characterization: structure, fabrication, electron density, and mobility. The measurements in magnetic fields parallel to the 2d plane correspond to the current being parallel to the magnetic field. The case in which the magnetic field was perpendicular to the current but parallel to the 2d plane was also measured. No anisotropy was observed.

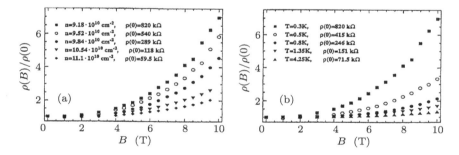

Fig. 4.13   Normalized resistance $\rho(B)/\rho(0)$ of a δ-doped GaAs/Al$_x$Ga$_{1-x}$As sample as a function of magnetic field B for different electron densities n at fixed temperature $T = 0.3$ K (a) and for different temperatures at fixed $n = 9.18 \times 10^{10}$ cm^{-2} (b) [18].

Figure 4.13(a) shows the resistivity $\rho(B)$ normalized by the zero-field resistivity $\rho(0)$ *vs.* the parallel magnetic field for several carrier concentration at $T = 0.3$ K. PMR is seen: the value $\rho(B)/\rho(0)$ increases with magnetic field, with the increase being stronger at lower electron densities. For carrier density $n = 9.18 \times 10^{10}$ cm^{-2}, the increase is almost sevenfold at a field of 10 T. Figure 4.13(b) shows MR at different temperatures for a fixed $n = 9.18 \times 10^{10}$ cm^{-2}. It is seen that the increase is stronger at lower temperatures. As noted, the magnetic field applied parallel to the single 2d conducting layer couples only to the electron spin. This implies that spins play an important role in 2d hopping conductivity and hints at the involvement of doubly-occupied states in the VRH process. The two electrons at the same site should have opposite spin. The second electrons charges the donors negatively, their binding energy with the donor is less than for the first electrons. This results in the formation of the "upper Habbard band" together with the usual impurity band with one electron at a donor (see [19] and Refs. therein). The localization length in the upper Habbard band is much larger than for the lower band, making the upper band dominant for the hopping process. Moreover, it can allow delocalization in principle, even if the states in the lower band are strongly localized. The magnetic field decreases the occupation numbers for the upper Habbard band due to the spin alignment, which leads to an increase of resistance, i.e., to positive MR. For 3d VRH conductivity, the corresponding mechanism was considered by Kurobe and Kamimura [20] and Agrinskaya *et al.* [21].

4.3 "Phononless" Hopping Conductivity

There is another manifestation of the influence of the spin interaction on 2d VRH conductivity, relating to a temperature-independent universal hopping prefactor

$$\rho_0 = \frac{h}{e^2} \qquad \text{or} \qquad \rho_0 = \frac{h}{2e^2}, \tag{4.7}$$

which has been reported for various 2d systems.

As an example, Fig. 4.14 shows the temperature dependence of resistivity for δ-doped GaAs/Al$_x$Ga$_{1-x}$As heterostructure at different electron densities reported by Khondaker *et al.* in Ref. [22]. At lower densities, all dependences fit straight lines when plotted as $\log[\rho(T)]$ *vs.* $T^{-1/2}$ with a common prefactor $\rho_0 = h/e^2$, while at higher densities, experimental data fit straight lines when plotted as $\log[\rho(T)]$ *vs.* $T^{-1/3}$ with a common prefactor $\rho_0 = 0.5h/e^2$. This is shown more clearly in Fig. 4.15, where the resistance in units $h/e^2 = 25.8$ kOhm is plotted *vs.* the normalized temperature. Furthermore, the temperature-independent prefactor is also material-independent, being the same for different materials with 2d hopping conductivity: Si-MOSFET [23], δ-doped GaAs [24] thin Be films [25]. This is in marked contrast to the predictions of the phonon-assisted mechanism, where the hopping prefactor must be temperature dependent

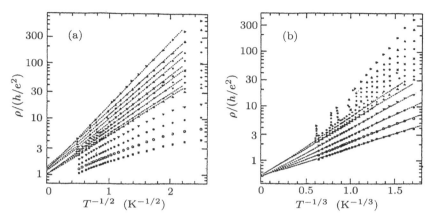

Fig. 4.14   The resistivity $\rho(T)$ normalized by h/e^2 vs. $T^{-1/2}$ — (a) and $T^{-1/3}$ — (b) for different electron densities n in a δ-doped GaAs/Al$_x$Ga$_{1-x}$As heterostructure [22]. The values of n from top to bottom are (in 10^{10} cm^{-2}) from 8.47 to 10.85.

and must be different for different host semiconductor material because it contains lattice parameters, including the velocity of sound, deformation potential and crystal density. Therefore, the observation of a material-independent and temperature-independent prefactor may be considered as evidence that hopping is not phonon-assisted ("phononless" hopping conductivity). One can assume that in this case hopping transition are assisted by some kind of electron–electron interaction (EEI-assisted VRH).

In contrast to these results, measurements of 2d VRH in heterostructures without δ-doping showed, that plotting the experimental data *vs.* $T^{-1/2}$ and $T^{-1/3}$ did not yield straight lines without the assumption that the hopping prefactor is temperature dependent in the form Refs. [26–28]:

$$\rho_0 = AT^m, \qquad m = 0.8 \div 1.0, \tag{4.8}$$

which is in agreement with the prediction of the conventional phonon-assisted hopping theory [9].

The possible reason for such a discrepancy will be discussed later, but now we point out the following fact: in some 2d systems, for electron densities near the metal–insulator transition, "phononless" (or EEI-assisted) VRH conductivity is observed. The origin of this hopping mechanism is not quite clear. It was suggested by Aleiner, Polyakov and Shklovskii [29] that the VRH conductivity may have a prefactor of the form $\rho_0 = (h/e^2)f(T/T_0)$, where $f(T/T_0) \approx 1$ for $T_0 \approx T$. However, there is no theoretical justification for this choice of prefactor in the framework of

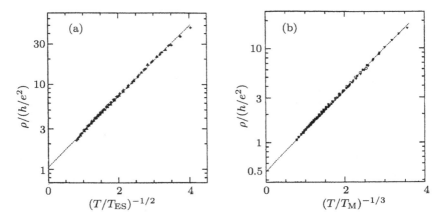

Fig. 4.15   Plots of the resistivity $\rho(T)$ normalized by h/e^2 *vs.* $(T/T_{\mathrm{ES}})^{-1/2}$ for carrier densities 9.18–9.52 × 10^{10} cm^{-2} — (a). The resistivity $\rho(T)$ normalized by h/e^2 *vs.* $(T/T_{\mathrm{M}})^{-1/3}$ in the Mott regime for carrier densities 9.84–10.85 × 10^{10} cm^{-2} — (b). The dashed line is a least-squares fit to the data with an intercept $\rho_0 = (0.49 \pm 0.02)h/e^2$. (Adopted from Ref. [22].)

the phonon-assisted model, which suggests another mechanism as discussed by Fleishman *et al.* [30].

Significantly, the prefactor in the ES VRH is twice that in the Mott VRH (Fig. 4.15). In the conventional theory, the conductivity prefactor is proportional to the DOS, and this change by a factor of two in the prefactor suggests that the DOS was altered by the lifting of the spin degeneracy.

The importance of the spin state of electrons in EEI-assisted hopping was established by the observation [18] that the EEI-assisted mechanism is destroyed in strong parallel magnetic fields and is replaced by conventional phonon-assisted hopping. In Fig. 4.16(a), $\log \rho_{xx}$ of the δ-doped GaAs/AlGaAs sample is plotted *vs.* $T^{-1/2}$ as in Fig. 4.14(a). In zero magnetic field, the data fit the straight line with the temperature-independent universal prefactor $\rho_0 = 25.8$ kOhm. However, in strong fields, the data show systematic deviation from the straight line. In this case, for analysis of the $\rho(T)$ dependences, it is useful to plot the derivative $w = -\partial[\ln \rho(T)]/\partial(\ln T) = p(T_0/T)^p$ as a function of T on a log–log scale, because the slope of the lines on this scale gives directly the value of p. The result of this analysis is shown in Fig. 4.17.

One can see that at zero field and low temperatures (below 1 K), the power p is indeed equal to 1/2, which corresponds to ES VRH. However, in strong magnetic fields, the value of p increases up to $p = 0.8$. There is

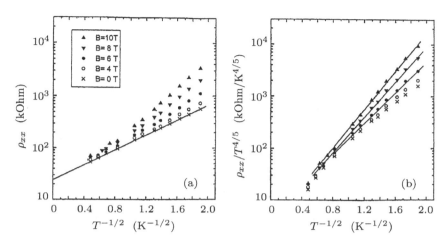

Fig. 4.16   Temperature dependence of ρ_{xx} (a) and $\rho_{xx}/T^{0.8}$ (b) *vs.* $T^{-1/2}$ for $n = 9.52 \times 10^{10}$ cm^{-2} at $B = 0$, 4, 6, 8 and 10 T [18]. The straight lines are guides to the eye.

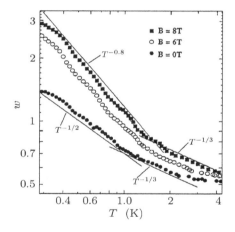

Fig. 4.17   The derivative $w = -\partial[\ln \rho(T)]/\partial(\ln T) = p(T_0/T)^p$ as a function of T on a log–log scale for $n = 9.52 \times 10^{10}$ cm^{-2} at $B = 0$, 6, and 8 T [18]. The exponent p is given by the slope of the curves.

no theoretical explanation for VRH conductivity with $p = 0.8$.

Therefore, one suspect that a stronger temperature dependence of the resistivity is caused by the temperature-dependent prefactor. Following Ref. [28], we assume that the prefactor becomes temperature dependent in the form of Eq. (4.8), that is $\rho_0 = AT^m$ with $m = 0.8$, and plot

$\log(\rho_{xx}/T^{0.8})$ *vs.* $T^{-1/2}$ in Fig. 4.16(b). It is seen that the low-temperature data for strong fields $B = 6$, 8, and 10 T fit the $T^{-1/2}$-law. This could be interpreted as the restoration of the conventional phonon-assisted hopping mechanism in strong parallel magnetic fields when electron spins are polarized.

A possible mechanism of EEI-assisted hopping has been proposed by Baranovskii *et al.* [31, 32]. They propose that the conductivity arises from resonant tunneling between transport states brought into resonance by Coulomb potentials produced by surrounding sites with fluctuating occupation numbers.

Before discussing this mechanism, two points must be recalled regarding the hopping theory.

(i) The energy level of each localized electron is fluctuating in time because it depends on the fluctuating occupation of surrounding localized sites by electrons. In theoretical consideration, the fluctuating energies were approximated by their average values.

(ii) The infinite percolation cluster through which the current flows contains many finite clusters in its pores (see Fig. 2.4, Chapter 2).

The states in the finite clusters do not participate directly in the current flow, but fluctuations in the site occupation lead to energy fluctuations for sites in the infinite cluster. This produces current hopping noise, which is a direct manifestation of the sequential Coulomb correlation in a hopping system, as first suggested by Knotek and Pollak [33, 34]. They proposed two types of electron–electron correlations: sequential correlations (SC) and electron polarons (EP). In the case of EP, several electrons hop simultaneously. Such simultaneous multi-electron transitions can have lower activation energies than single-electron hops and therefore can be efficient at low temperatures. In the case of SC, the single-electron current-determining hops can be assisted by preliminary hops between surrounding states. These preliminary hops prepare the favorable situation for the current-carrying transitions of single electrons.

The mechanism suggested in Refs. [31, 32] is based on the assumption that the current-carrying hop between sites belonging to the percolation cluster is due to a phononless resonant tunneling of a single electron between occupied and empty sites, with the resonance being prepared by fast assisting hops in the environment of these sites. In this mechanism, two neighboring sites i and j belonging to the percolation cluster are considered, so that an electron transition between them is necessary for DC conductivity. It is assumed that fast electron transitions in the sites surrounding i and j, cause the electron energies ε_i and ε_j on the chosen sites to fluctuate with some typical frequency ω due to the Coulomb potentials of surrounding electrons. In fact, it is sufficient to consider the fluctuations of either energies ε_i or ε_j. Due to this electron correlations, a resonant situation

$$|\varepsilon_i - \varepsilon_j| \leq I_{ij} \tag{4.9}$$

can be established for some short time interval t. Here $I_{ij} \propto \exp(-2r_{ij}/a)$ is the overlap integral between sites i and j. If the amplitude of the energy fluctuations is $E \gg I_{ij}$, this time interval can be approximated by

$$t \approx \frac{I_{ij}}{E\omega}. \tag{4.10}$$

The spreading of the electron wave function between the resonant sites obeying condition (4.9) takes a time τ of order $\hbar/I_{ij}$. If the energy fluctuations are fast compared to τ, i.e., $\omega\tau \gg 1$, then the probability of the transition occurring during one fluctuation period is t/τ and the transition rate is

$$W \approx \frac{\omega t}{\tau}. \tag{4.11}$$

Using Eq. (4.10), one obtains

$$W \approx \frac{I_{ij}^2}{E\hbar}. \tag{4.12}$$

This formula shows that the frequency ω of the assisting hops is not involved in the hopping rate W for the current-carrying hops if $\omega\tau \gg 1$, i.e., if the assisting hops are much faster than the current-carrying hop. This condition is the main requirement of all SC models.

In the model discussed, the current-carrying pair of sites i and j must have only one electron. This is determined by the energy distribution function of the electrons $f(\omega)$. Therefore, the transition rate ν_{ij} from the site i to the site j has the form

$$\nu_{ij} = W f(\varepsilon_i)[1 - f(\varepsilon_j)]. \tag{4.13}$$

In the Ohmic regime, $f(\varepsilon)$ should correspond to the Fermi distribution. This leads to the known expression

$$\nu_{ij} = \nu_0 \exp\left(-\frac{2r_{ij}}{a}\right) \exp\left(-\frac{\varepsilon_{ij}}{T}\right), \tag{4.14}$$

which is similar to the conventional expression considered in phonon-assisted hopping theory (see Eq. (2.2) in Chapter 2), with the only difference in the origin of prefactor. To calculate the VRH conductivity using "phononless" transition rates determined by Eqs. (4.13) and (4.14), one must repeat the standard derivation procedure described in Chapter 2. This automatically leads to expression (4.1) for the VRH conductivity.

The suggested model of electron transport connects together a resonant tunneling mechanism of band conductivity characterized by the term e^2/h with the necessity of finding an occupied initial state and an empty final state. This results in the temperature dependence of the band-like electron motion in the form of the Mott or the ES VRH regime. One may refer to this mechanism as Variable-Range Resonant Tunneling (VRRT).

The VRRT mechanism does not seem to be connected with the state of electron spins. Therefore, it is not clear why EEI-assisted hopping is destroyed in strong parallel fields and is replaced by a conventional phonon-assisted mechanism (Fig. 4.16). To understand this fact, one should remember that in VRH, tunneling occurs over a long distance, much longer than the average distance between impurities. We have already mentioned the Spivak model [15] in which the probability of a hopping transition over long distances depends on the mutual spin states of the tunneling electron and an electron localized on the nearby occupied state (Fig. 4.10). Similarly, resonant tunneling may also be suppressed by the spin alignment of electrons.

The other possible mechanism of "phononless" hopping is based on the existence of delocalized tightly bound electron–hole pairs (excitons) in the system as suggested by Berkovits and Shklovskii [35]. In this model, the behavior of the many-level spectral statistics in the strongly disordered regime is explained as a manifestation of exciton delocalization, which gives rise to the idea that delocalized excitons may replace the phonons as the energy source for the hopping conductance.

We have to admit that a self-consistent many-body theory of VRH is still lacking and, therefore, the conditions that favor EP or SC in the EEI-dominant regime are not clear. Moreover, it is also not yet clear under which experimental circumstances VRH is dominated by EEI mechanism. The problem is to find the conditions under which phononless VRRT becomes more efficient than phonon-assisted hopping. For example, in the experiments of Khondaker *et al.* [22], a universal prefactor was observed, whereas Keuls *et al.* [28] reported the conventional phonon-assisted mechanism for 2d VRH in a similar GaAs/AlGaAs heterostructure. The only difference was the presence of the δ-doped layer near the 2d conducting plane in Ref. [22]. This suggests that the existence of a δ-doped layer is a significant factor for the dominance of EEI-assisted hopping. To check this idea, the authors of Ref. [24] reviewed the results of measurements of the low-temperature conductivity in a gated GaAs sample [36]. This sample contained two parallel δ-doped layers whose carrier concentration can be varied by the gate voltage V_G. This allows one to alter the conductivity of the layers from almost metallic to insulating behavior. Details of the sample structure, as well as measurement technique, can be found in Ref. [36].

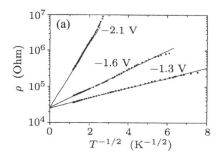 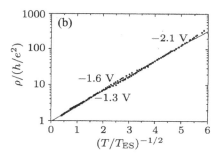

Fig. 4.18 Temperature dependence of the resistivity of a δ-doped GaAs sample on a scale lg ρ vs. $T^{-1/2}$ for three values of the gate voltage — (a). The same data plotted in dimensionless units $\rho/(h/e^2)$ vs. $(T/T_{ES})^{-1/2}$ — (b)[24].

Figure 4.18(a) shows the temperature dependence of the sample resistance at different gate voltages plotted as a function of $T^{-1/2}$. The experimental data fit a straight line on such a plot. The slope of straight lines gives the value of $T_0 \equiv T_{ES}$, while the intersection with the ordinate axis gives the prefactor ρ_0. One sees that all three straight lines have the same prefactor, which is close to the quantum of 2d conductivity $h/e^2 = 25.8$ kOhm. This universality is shown more clearly in Fig. 4.18(b), where the temperature dependences of the resistivity are plotted in dimensionless units $\ln[\rho/(h/e^2)]$ vs. $(T/T_{ES})^{-1/2}$. It is seen that the ES VRH is valid over a wide interval of resistances and temperatures with the universal prefactor $\rho_0 = (h/e^2)$.

This result supports the assumption that the presence of charged impurities (ionized donors) directly in the conducting 2d plane is favorable for the EEI-assisted mechanism of hopping. It is worth mentioning that the inversed layer in Si-MOSFET can also be characterized by a high concentration of charged surface defects and impurities, situated directly in the 2d conducting plane. It is likely that the presence of charged centers so close to the conducting path produces very pronounced dynamical fluctuations of the site energies, which is necessary for the VRRT mechanism of conductivity. However, this idea requires further theoretical study.

In conclusion, it must be admitted that in spite of many efforts to understand the experimental observation of spin-dependent VRH and 2d hopping conductivity with the universal prefactor, a rigorous theory is not yet available and many problems discussed above are still unresolved.

Chapter 5

$1/f$ Noise; Application of the Hopping Conduction

5.1 Introduction

In many conducting materials, low frequency current noise is characterized by a well-known $1/f$ form [1, 2] whose spectral density S is proportional to the square of the applied voltage V and inversely proportional to the frequency f (Hooge formula):

$$S = \frac{\alpha V^2}{n\Omega f}. \tag{5.1}$$

Here α is a numerical coefficient (Hooge parameter), n is the carrier concentration, and Ω is the sample volume. $1/f$ noise is particularly large in inhomogeneous conductors which are characterized by the high density of electron traps having a wide spread of relaxation times. Therefore, it is interesting to study $1/f$ noise in hopping conductivity where the DC current flows via a critical network of donors with a very wide distribution of the local hopping probability w_{ij} between two localized states i and j:

$$w_{ij} = w_0 \exp(\xi_{ij}), \qquad \xi_{ij} = \frac{2r_{ij}}{a} + \frac{\varepsilon_{ij}}{T}, \tag{5.2}$$

where ε_j, ε_i are the energies of these levels, $\varepsilon_{ij} = \varepsilon_j - \varepsilon_i$, r_{ij} is the distance between the initial and final states, a is the Bohr radius.

In an early investigation of the low-frequency noise in the silicon inversion layer, Voss found that $1/f$ noise is an intrinsic property of hopping conduction [3]. Substantial interest in the investigation of the hopping noise originates also due to the fact that low-frequency $1/f$ noise limits the characteristics of devices working in the hopping regime, such as thermoresistors (thermistors) used for measurements of low and superlow temperatures (well below 1 K). Thermistors are also used in high-sensitive bolometers, detectors of far-infrared radiation, microcalorimeters in astrophysics for X-ray spectroscopy (see, for example, Refs. [4–6] and references therein).

5.2   Low-frequency $1/f$ Hopping Noise

In the regime of hopping conductivity, the DC current flows via so-called "infinite percolation cluster" or "critical network" (CN) which consists of the random network of resistances

$$R_{ij} = R_0 \exp(\xi_{ij}), \qquad \xi_{ij} \leq \xi_{\mathrm{c}}, \qquad (5.3)$$

where ξ_{c} is the maximal value of ξ_{ij} which preserves the network connectivity. There are two regimes of hopping transport:

(i) nearest-neighbor hopping (NNH) is realized at relatively high temperatures, where only the spatial distribution of localized states (for concreteness, donors for n-type semiconductor) determines the formation of the percolation cluster. Fluctuations in energy ε_{ij} play a secondary role and do not influence the formation of CN because the first term in ξ_{ij} in Eq. (5.2) is much larger than the second one. This leads to a conductivity with constant energy of activation ε_3, with temperature-independent average hopping distance r_{c} and constant CN;

(ii) at low enough temperatures, the second term in Eq. (5.2) becomes comparable with the first one, and the NNH regime is replaced by the variable-range hopping conductivity (VRH) where the CN is formed by taking into account the nearness of the energies ε_j, ε_i to the Fermi level. In the VRH regime, the average distance of hopping $r(T)$, as well as the typical size of the cells L in CN continuously increases with decreasing temperature T (see Fig. 2.4 in Chapter 2).

There are only two parameters which could contribute the fluctuation of the sample conductance: the total number of charge carriers (concentration n) and their mobility μ. For hopping conductivity, the "mobility" is unusually small and cannot be described in the framework of the Boltzmann equation. In the hopping regime, the probability w_{ij} of a tunneling transition between two localized levels, Eq. (5.2), plays the role of the "mobility".

The low-frequency noise in the NNH regime was first studied theoretically by Shklovskii [7], who suggested that the noise is caused by fluctuations in the carrier number ("n-fluctuations") due to electron exchange between donors in the critical network (CN) and other donors in its interstices (Fig. 5.1). By definition, donors not in CN are separated from any donor in that network by distances longer than the maximal distance in CN.

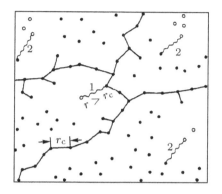

Fig. 5.1 A fragment of the critical network (CN) for NNH regime [7]: conducting path (solid lines) and porous are shown. The maximal distance between donors in CN is r_c. All distances in "fluctuators" $r > r_c$. 1 — electron traps which produce an "n-noise", 2 — "μ-noise" fluctuators.

Thus electron exchange between such separate donors and CN-donors takes place very slowly, which leads to excess noise at low frequencies. This idea is similar to the McWorter explanation of the $1/f$ noise in MOSFET's in terms of the electron exchange between the conducting 2d electron gas and electron traps in the oxide with an exponential wide distribution of tunneling times [8].

The investigation of hopping noise in amorphous silicon films [9] showed that the hopping mechanism does indeed give rise to noise with $1/f$-dependence at low frequencies, but a quantitative estimate of the noise parameters using expression (5.1) was questionable because of the uncertainty in the definition of the carrier concentration in amorphous materials. To avoid this difficulty, the low-frequency noise was measured in Ref. [10] in samples of crystalline n-Ge:As and p-Ge:Ga having an exactly known concentration of charge carriers.

Figure 5.2 shows the temperature dependence of resistivity for three samples: n-Ge:As (1, 2) and p-Ge:Ga (3). The concentration of carriers at 300 K for n-Ge:As was $n = 7.5 \times 10^{16}$ cm^{-3} and 1.2×10^{17} cm^{-3} (1, 2) and $p = 4.6 \times 10^{16}$ cm^{-3} for p-Ge:Ga (3), compensation $K = 0.01$. Below 3 K, the temperature dependence of resistivity $\rho(T)$ is seen to obey the NNH law with constant energy of activation ε_3. Figure 5.3 shows the noise spectral density S for the p-Ge:Ga sample measured at different applied electric fields at a fixed temperature of 2.5 K. One can see that S is indeed proportional to V^2/f as suggested in Eq. (5.1). This allows one to use the parameter S/V^2 at fixed frequency (for example, at $f = 10$ Hz) to characterize the magnitude of the noise.

Figure 5.4 shows the temperature dependence of this quantity for all three samples. It was found surprisingly that these dependences are much weaker than the temperature dependence of the resistances. This is the central experimental result of this investigation.

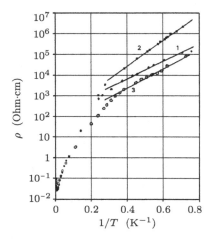

Fig. 5.2 Temperature dependence of resistivity of the samples n-Ge:As (1, 2) and p-Ge:Ga (3)[10]. The straight lines are a guide to the eye.

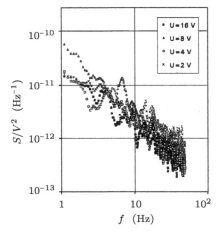

Fig. 5.3 Frequency dependence of the reduced voltage noise S/V^2 of the p-Ge:Ga sample at $T = 2.5$ K and different applied voltages on log–log scale [10].

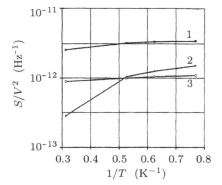

Fig. 5.4 Temperature dependence of the noise spectral density at $f = 10$ Hz for samples: 1, 2 — n-Ge:As, 3 — p-Ge:Ga.

The "concentration" model of hopping noise, developed by Shklovskii [7], predicts saturation in the noise at low frequencies which has not yet been seen. In the alternative "mobility" approach suggested by Kozub [11], resistance fluctuations come from the change in the value of ε_{ij}. In NNH conductivity, all donors belonging to CN are fixed. Therefore, all hopping distances r_{ij} on the percolation level are constant and fluctuations in w_{ij} are only due to instabilities in the energies ε_{ij} of the impurity levels. These fluctuations are caused by the random electric field generated by the states outside the CN due to random occupancy of the states. There are some pairs of sites separated from the percolation cluster and characterized by wide scatter of the characteristic hopping time. These pairs are, in fact, "electronic fluctuators". Note that this effect has already been discussed in Chapter 4 as the explanation of electron–electron interaction-assisted ("phononless") hopping conductivity.

In the NNH mechanism, where the CN is constant and fluctuations $\delta\varepsilon_{ij}$ cannot change the structure of the infinite cluster, the resistance fluctuations are caused by fluctuations in the value of the maximal resistance R_c (with $\xi = \xi_c$) in CN. The situation is different in VRH, because the value $\varepsilon_{ij} = \varepsilon_j - \varepsilon_i$ affects the formation of CN. At low temperatures, if the fluctuation of this value $\delta\varepsilon_{ij} > T$, this could lead to a change of the shape of the percolation cluster; this would remove from CN the resistors with $\xi_c > \xi_{ij} > \xi_c - 1$ or, alternatively, introduce some other resistors into CN. The "mobility" approach to hopping noise predicts that the spectral density of noise is weakly-temperature dependent in the NNH regime (this agrees with the experimental observation, see Fig. 5.4) and increases rapidly with decreasing T in the VRH regime, as $T^{-3/2}$ for the Mott VRH and T^{-3} for the ES VRH [11].

The increase of the noise results from the decrease of the density of the CN, because with decreasing T, fewer and fewer states with energies close to the Fermi level are involved in the formation of the CN which becomes thinner. This agrees with a prediction of the Hooge relation (5.1) of an increase of noise with a decrease of the "noisy volume".

Thus, there are two models for explaining the $1/f$ hopping noise, the Shklovskii approach of concentration fluctuations ("n-noise") [7] and the Kozub approach of "mobility" fluctuations ("μ-noise") [11]. The "n-noise" model [7] is based on a slow exchange of electrons between a donor belonging to the infinite cluster where the distances between donors satisfy $r \leq r_c$ and an insulated donor located in a spherical pore of large radius $r > r_c$ which serves as an "electron trap". The "μ-noise" model [11] considered

pairs of donors ("fluctuators") both being outside the conducting percolation cluster, slowly exchanging electron with each other and modulating
transport on current paths by their fluctuating potential. At first glance,
these two approaches are different. But this is not so. In subsequent work
[12], Shklovskii showed, that an exchange of an electron between a donor
belonging to CN and "electron trap" also forms a "fluctuator". Capture
and release of electrons results not only in a fluctuation of the number of
conducting electrons ("n-noise"), but also to a fluctuation of the hopping
probability ("μ-noise") due to a variation of the trap potential. Uniting
theoretical consideration [12] lead to the following expression for the conductivity fluctuations:

$$\frac{I_f^2}{I^2} = \frac{\alpha(f,T)}{f N_D V},\tag{5.4}$$

which is, in fact, the Hooge law (5.1). Here, $(I_f)^2$ is the spectral density of
current fluctuations and I is the average current through the sample, $N_D V$
is the total numbers of donors and α is the Hooge coefficient:

$$\alpha(f,T) = \exp\left[-\frac{\pi}{6}N_D a^3 \ln^3\left(\frac{w_0}{f}\right)\right]\exp\left(\frac{T_c}{T}\right)^3.\tag{5.5}$$

Here, $T_c \approx (e^2/\kappa)a N_D^{2/3}$ is the critical temperature of the transition from
NNH to VRH (see Eq. (2.15) in Chapter 2). Equation (5.5) is valid both
for the NNH ($T > T_c$) and for the VRH ($T < T_c$). The first exponential
factor reflects the probability of a pore in the percolation cluster free of
donors (traps) which are able to capture and release an electron with an
exchange rate greater than f. This factor is temperature independent. The
second exponential factor depends on temperature, but for the NNH regime
in which $T \gg T_c$, this term is close to unity. This makes the temperature
dependence of the noise amplitude very weak, in agreement with experiment
(Fig. 5.4).

In the VRH regime, the exponential growth of $1/f$ noise with decreasing
T agrees qualitatively with measurements for implantation doped silicon
samples [4], but disagrees with results obtained on bulk silicon crystals [13].
$1/f$ noise was also studied in two-dimensional n-GaAs channels [14] over
a wide range of temperatures including the transition between NNH and
VRH. In agreement with Eq. (5.5), it was observed that the noise amplitude
decreases sharply with increasing temperature in VRH range, while in NNH
range the temperature dependence becomes weak. Thus, in the NNH range,
the relative intensity of $1/f$ noise is almost temperature independent in
spite of the activation temperature dependence of the conductivity.

In the hopping regime, the exact value of the Hooge numerical coefficient α is difficult to determine because of the complex dependence of conductivity (or current I) on the charge carrier concentration n. In MOSFET's in which most of the electrons are free, fluctuation in n obviously lead to proportional fluctuations in conductivity. In hopping conductivity, in the NNH regime, this relation is not so obvious. Only in the case of strongly compensated samples, for which the concentration of minority impurities N_A (acceptors in n-type semiconductors) is almost equal to the concentration of donors N_D, the value of hopping conductivity is determined by the density of remaining electrons $n = N_D - N_A$. For the case of weakly compensated samples $K = N_A/N_D \ll 1$, the role of charge carriers is played by the small number of empty donors (holes). Let us estimate the value of α for sample 3 p-Ge:Ga (Fig. 5.3). For this sample, the concentration of majority impurities (acceptors) was $N_A = 4.6 \times 10^{16}$ cm^{-3}, $K \equiv N_D/N_A \approx 0.01$, the total number of charge carriers for hopping conductivity is therefore $N_A K \approx 4.6 \times 10^{14}$ cm^{-3}. For a sample volume $V = 2.5 \times 10^{-3}$ cm^3, one obtain $\alpha \approx 3.5$.

Even for weak and strong compensation in NNH regime, the relation between hopping conductivity and the concentration of charge carriers is not quite linear. For intermediate compensation $K \approx 0.5$, the proportionality between n and I is obviously non-linear. For VRH conductivity, the connection between fluctuations in n and I is even more complex. For example, any change of n due to fluctuation usually leads to a corresponding shift in the position of the Fermi level (FL). However, for ES VRH, when the Coulomb gap exists in the density of state at the FL, the Coulomb gap moves together with the FL and to first approximation, the density-of-state remains unchanged at small energies around the FL. This means that there is no influence of the variation of n on the current in the ES VRH regime and, therefore, it is unclear whether the "n-noise" mechanism and Hooge law is valid and how to calculate the parameter α in Eq. (5.1). One can suggest that in this case, resistance fluctuations are determined by the "μ-noise" mechanism.

A new model of the hopping noise was later suggested [15], in which $1/f$ noise is related to many-electron transitions of the chessboard clusters, whose rate decreases exponentially with cluster size. The slow fluctuations of cluster states modulate the critical resistors forming the conductive backbone cluster leading to noise in electronic conductivity.

It follows from the above consideration, that the $1/f$ noise is well established experimentally in the different regimes of hopping conductivity, but a complete theory of the low-frequency hopping noise is still not available.

5.3   Devices Based on Hopping Conductivity

The most important devices where the hopping conductivity plays a crucial role are low-temperature thermoresistors ("thermistors"). These devices have been used for many years as detectors of different kind of electromagnetic irradiation, as sensitive bolometers for infrared astronomy, for simple and low-inertial thermometers for low (below 4 K) and superlow (well below 1 K) temperatures. The detailed survey of this issue can be found in the review of McCammon [16].

Using doped germanium and silicon as thermistors has a long history. At low temperatures, thermocouples and metallic resistance thermometers rapidly lose sensitivity. The semiconductor thermometers available that time were mostly metal oxides, whose resistance increased rapidly when cooled and soon became immeasurable. In the early 1950s, a doped germanium sample was used as a suitable thermometer at temperatures as low as 1 K [17].

Later, the practical Ge:Ga low-noise devices were developed which operated at liquid helium temperature [18]. A thermistors with a hopping mechanism of conductivity could also be used as an efficient radiation absorber ("bolometer"). Therefore, germanium bolometers were soon applied to the rapidly grown field of infrared astronomy.

Today, the well developed technology of doping Ge and Si allows one to fabricate thermistors and bolometers as individual devices and in array of identical pixels together with all necessary interconnections. This success is based on a deeper understanding the main features of hopping conductivity discussed in the previous chapter: the temperature dependence of resistance in NNH and VRH conductivity, the transition from NNH to the Mott VRH, the crossover from the Mott VRH to ES VRH, the appearance of the "hard magnetic gap" at very low temperatures, smearing of the Coulomb gap with increasing temperature, non-linearity of the current–voltage characteristics, and positive and negative magnetoresistance which shows the sensitivity of the thermometer indication to the value of magnetic fields.

The temperature dependence of the resistance $R(T)$ for most thermistors in low-temperature range $(4.2 \div 0.3$ K) obeys the ES VRH law,

$$R(T) = R_0 \exp\left(\frac{T_0}{T}\right)^{1/2}. \tag{5.6}$$

For example, Fig. 5.5 shows $R(T)$ for six thermistors made from ion-implanted Si [19]. One sees that the "$T^{-1/2}$-law" is followed over several orders of magnitude in resistance. However, at "superlow" temperatures, systematic deviations from this behavior are observed: each curve deviates upwards from the straight line $\log R \propto T^{-1/2}$ to a stronger temperature dependence. These deviations are more clearly seen if one plot the ratio of the real data $R(T)$ to the straight-line fit as a function of dimensionless temperature T_0/T (Fig. 5.6). The reason of such deviations has already been discussed in Chapter 4. It was shown that this effect is determined by a transition to the "$1/T$-law" due to the appearance of a "magnetic hard gap" caused by the electron–electron spin exchange interaction.

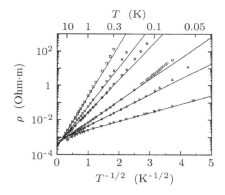

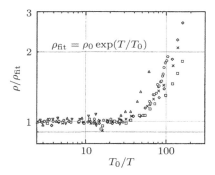

Fig. 5.5 Resistivity of ion-implanted Si:P, B samples [19]. The straight lines are fits to the ES VRH model valid over a temperature range $6.5 < T_0/T < 24$.

Fig. 5.6 Resistivity of samples in Fig. 5.5 divided by the ES VRH model as a function of normalized inversed temperature T_0/T [19].

In practice, it is useful to have an analytic expression for $R(T)$ over the entire temperature range. This is desirable for two reasons. First, it is convenient for thermometric purposes, since only a small number of calibration points are required to fix the entire $R(T)$ dependence. Second, at "superlow" temperatures (well below 1 K), indication of a thermistor can be influenced by some external heating due, for example, to radio-frequency noise. The parameters of the analytic function $R(T)$ have to be fit to measurements at higher temperatures where these effect are usually negligible and then extrapolated down to the lowest temperatures in order to compare with thermometer indication. This will allow one to detect any heating problems.

To operate at superlow temperatures, the samples must be doped very close to the critical point of the metal–insulator transition n_c, because samples with $n < n_c$ become highly resistive at millikelvin temperatures and practically unmeasurable, while samples with $n > n_c$ become metallic and lose their temperature sensitivity. The lower the working temperature interval, the closer should be the concentration of the carriers n to the critical value n_c. This requires a very precise and homogeneous method of doping. For Ge, extremely reproducible thermistors can be obtained from the NTD (neutron transmutation doping) method which consists of irradiation pure Ge crystals by reactor neutrons (see Chapter 2, Eq. (2.28)). Upon capturing a thermal neutron, three of five stable isotopes of Ge: ^{70}Ge, ^{74}Ge, and ^{76}Ge transform to ^{71}Ga (shallow acceptor in Ge), ^{75}As (shallow donor), and ^{77}Se (deep donor). The concentration of the NTD-introduced impurities N_{NTD} (cm^{-3}) is linearly proportional to the integral dose of neutron irradiation Φ (cm^2):

$$N_{\text{NTD}} = \alpha\Phi, \qquad \alpha = N_{\text{Ge}}x\sigma_x, \tag{5.7}$$

where α (cm^{-1}) is the reaction coefficient, N_{Ge} is the density of Ge atoms in the lattice, x is the portion of a given isotope in natural Ge, and σ_x is the cross-section for thermal neutron capture for the given isotope σ_x measured in barns (1 barn $= 10^{-24}$ cm^2).

The NTD of intrinsic Ge yields a series of p-Ge samples doped with Ga and compensated by As and Se. The degree of compensation K is about 32–40 %. The scatter in K arises from the fact that transmutation doping also involves epithermal neutrons, therefore the real values of cross-section differs somewhat from the values σ_x tabulated for thermal neutrons. As a result, the value of α varies slightly depending on the actual energy spectrum of neutrons in the reactor used for irradiation. It is also possible to obtain n-type Ge samples using the NTD method. For this purpose, the initial Ge crystal must be highly enriched with isotope ^{74}Ge, which transmutes into a As donor.

The NTD process does not end by the irradiation of samples or ingots in a nuclear reactor. The presence in the reactor spectrum of fast, energetic neutrons gives rise to the creation of radiation defects or even disordered amorphous regions in the irradiated sample. Defect annealing is a complex technological problem, because radiation defects cause the formation of complexes with the impurities contained in the starting material. This requires different annealing regimes for different materials having different contents of some deep residual impurities (oxygen, carbon). The importance of annealing is shown in Fig. 5.7, where the dependence of the free-electron concentration in NTD-^{74}Ge n is plotted *vs.* the thermal neutron irradiation dose Φ under different annealing conditions [20, 21]. Note the saturation of the linear dependence $n(\Phi)$ at very high doses Φ. The fact is that such high doses are achieved for a long time of exposure (in this experiment, more than 11 days) which is much larger than the half-life of transmutation of isotope ^{75}Ge in ^{75}As. Therefore, long time of exposure leads to formation of complexes of radiation defects with As atoms which prevent As atoms to be donors and impede the annealing of radiation damage. However, at low and moderate time of irradiation, the linear dependence $n(\Phi)$ holds with high accuracy.

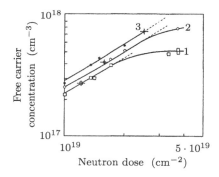

Fig. 5.7 Free-electron concentration in NTD-^{74}Ge *vs.* thermal neutron irradiation dose measured after annealing at 460 °C for 24 h (1), 50 h (2) and 100 h (3) [20, 21].

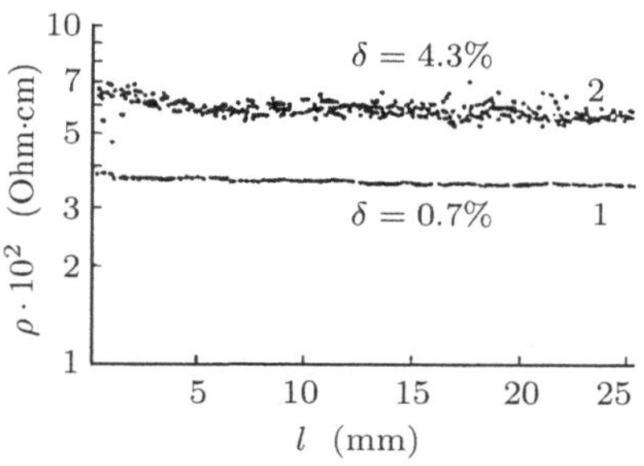

Fig. 5.8 Microdistribution of resistivity in Ge low-resistance standard samples doped by NTD (1) and by conventional metallurgical melt growth method (2) [22].

The NTD method has two advantages.

(i) The concentration of the induced impurities for constant neutron flux is proportional to the time of irradiation and can be controlled and reproduced with high accuracy.

(ii) Since the isotopes are chemically identical, they are randomly and uniformly distributed in Ge lattice. The cross-section of the neutron capture is relatively small for all isotopes. Therefore, transmutation events are rare and impurities are introduced quite uniformly.

Figure 5.8 shows the resistance inhomogeneity in a high quality Ge metallurgically doped sample used as a standard resistor and in a regular NTD-Ge sample [22]. In the latter case, one sees that the inhomogeneity is extremely small, even below the measurement accuracy.

Doping homogeneity through large blocks of Ge allows one to fabricate many very uniform thermometers which is important for the construction of the large arrays and a matrix of detectors for image processing in the far-infrared region. Figure 5.9 shows the coincidence of the calibration characteristics for three thermistors made from one NTD-Ge plate [23].

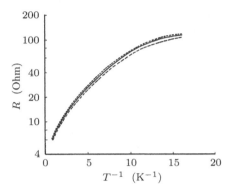

Fig. 5.9   Calibration characteristics of three thermistors made from one plate of NTD-Ge [23].

However, the NTD method has obvious disadvantage. The high penetrating ability of the neutron flux, which provides homogeneity of doping, also means that it is impossible to protect a part of the sample from neutrons and to dope only in selected areas in a crystal. As a result, all elements in a matrix of detectors must to be fabricated and mounted individually. This is a drawback for integrated circuit electronics (mainly based on silicon) where detectors are fabricated on one wafer together with other elements of the circuit and interconnections.

For Si, the coefficient of NTD doping α in Eq. (5.7) is small, two orders of magnitude less than for Ge. Moreover, the critical concentration of the metal–insulator transition in Si is one order of magnitude larger than in Ge. As a result, for Si, the NTD method cannot be used for fabrication of low-temperature thermistors because it requires an extremely large dose of neutron irradiation. Instead, doping with implanting ions is used for this purpose. This technique is well-developed in semiconductor electronics, and it provides a penetration depth of up to 1 μm, depending on the energy of the implanting ions. The depth distribution of the ions obeys the Gaussian profile. However, a more uniform distribution and almost flat-top profile can be achieved by irradiation with different energies of ions with controlled doses, followed by treating at high temperatures to allow the diffusion of the implanted ions. The obvious advantage of this doping method is that it can be masked by standard photolithographic technique which allows the simultaneous fabrication of a large numbers of small-size thermistors with fully integrated electrical connection. However, in contrast with NTD-Ge thermistors, the ion implanted Si thermistors suffer from a lack of reproducibility in spite of closely controlled energy and doses of the implanted ions.

One concludes that doped semiconductors with VRH conductivity, such as NTD-Ge or ion-implanted Si, are excellent materials for low-temperature thermistors. The main parameter of a thermistor is the sensitivity

$$\beta = \frac{d\log(R)}{d\log(T)},$$ (5.8)

which, in the case of ES VRH gives $\beta = 0.5(T_0/T)^{1/2}$. One can control T_0 by the doping level, making β as large as desired. It is also possible to fabricate thermistors with a small size to make the heat capacitance negligible, in spite of the relatively high specific heat of the doped semiconductors. However, a decrease of size leads to an increase of electric field at fixed voltage, which in turn leads to non-linearity of the current–voltage characteristics. This effect has to be taken into account in designing the optimal size and the value of T_0 for a given application.

Phonon-assisted hopping is an inherently non-linear process, so Ohm's law is valid only in the limit of small fields E, when the energy difference on the hopping length $eEr \ll T$. For intermediate electric fields, the theory predicts an exponential decrease of resistance with increasing E (see Eq. (2.39)):

$$R(T,E) = R(T,0)\exp\left(-\frac{eEr}{T}\right).$$ (5.9)

Here $R(T,0)$ is the resistance in the limit of low electric fields. In addition, the mean hopping distance $r(T)$ increases with decreasing temperature in the VRH regime. For ES VRH, $r(T) = (a/2)(T_0/T)^{1/2}$. This leads to [24]

$$R(T,E) = R(T,0)\exp\left(-\frac{eEaT_0^{1/2}}{2T^{3/2}}\right).$$ (5.10)

It follows from Eq. (5.10) that for a given thermistor with fixed T_0, the data plotted as $\log R$ *vs.* $E/T^{3/2}$ for different lattice temperatures (or the "heat sink temperature" T_s) should be parallel straight lines of the equal slope. Experimental data obtained for a NTD-Ge thermistor [24] are in agreement with this prediction (Fig. 5.10).

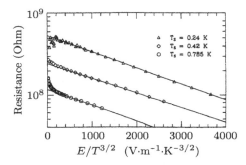

Fig. 5.10 Resistance *vs.* $E/T^{3/2}$ for a NTD-Ge sample with $T_0 = 59.4$ K at three lattice temperatures T_s [24].

An additional source of the decrease of resistance with increasing applied voltage is the current induced heating of electrons. In conventional conductivity, the applied power I^2R is absorbed by the free electrons and transferred to the lattice via electron–phonon interactions. At superlow temperatures, the electron–phonon coupling is very weak, and the energy is distributed among electrons faster than it can be transferred to the lattice. Therefore, the "electron temperature" T_e is higher than the lattice temperature (or "heat sink" temperature T_s). Free electrons in metals accelerate in an electric field and, therefore, take the energy directly from the electric field. For localized electrons, this is impossible and "heating" implies a change in the distribution function: the probability for electrons to occupy states of high energy increases in strong electric field, which can be interpreted also as $T_e > T_s$. The sample resistance is determined by T_e which therefore must be included in Eqs. (5.9) and (5.10).

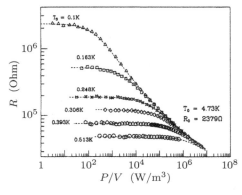

Fig. 5.11 Resistance of an ion-implanted Si device with $T_0 = 4.73$ K, and $R_0 = 2.379$ Ohm plotted *vs.* power density P/V for various lattice temperatures [24].

Figure 5.11 shows the dependences of the resistance of an ion-implanted Si thermistor as a function of the power density P/V measured at different lattice temperature T_s [24]. One sees that for T_s about 0.5 K, the resistance remains constant up to relatively high power densities. However, with decreasing T_s, R start to decrease at lower P/V, and finally, at T_s about 0.1 K, there is almost no interval of power density over which the resistance is constant.

In summary, both the electric-field effect and the power-density effect lead to a decrease of R. However, it is difficult to distinguish between these two effects because they are related to each other: $P/V = E^2/\rho$, where ρ is the sample resistivity.

The choice between NTD-Ge and ion-implanted Si thermistors depends on the application: excellent reproducibility of the NTD-Ge *vs.* easy fabrication of large number of small integrated devices in the case of ion-implanted Si. However, it is interesting to compare their universal characteristics such as power handling capability per unit heat capacity at the same thermistor sensitivity β. The last requirement implies comparing two devices with approximately equal T_0. Figure 5.12 shows the reduction in resistance as a function of power density for two thermistors made from NTD-Ge and ion-mplanted Si with almost identical T_0 [24]. The Ge data are multiplied by 50. It means that Si and Ge thermistors have the same non-linearity, but the Ge-device operates at 50 times lower power density. It was noted [24] that the doping density in Ge is also 50 times smaller than in Si. Therefore, if the curves were plotted in terms of Watts per impurity atom instead of Watts per m^3, they would merge without shifting.

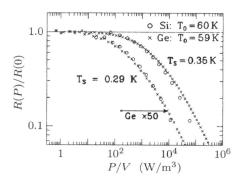

Fig. 5.12 Resistance as a function of power density at two lattice temperatures for Ge and Si thermistors with similar temperature sensitivity. The Ge data are multiplied by the factor of 50 [24].

The non-ohmic behavior induced by electric field effect or power per volume limitation, is a drawback for semiconductor thermistors, because cryogenic calorimeters require a very small volume to make the heat capacity negligible. This means a tradeoff of heat capacity for sensitivity. However, this shortcoming can occasionally be converted into an advantage. The hot electron effect allows the simple thermometer to act as a universal detector of heating, where the electron system is both the absorber and the thermometer, and the thermal isolation is provided by strongly reduced electron–phonon coupling [16].

The non-ohmic behavior also limits the only significant thermometer parameter — the sensitivity β (5.8). Hot electrons and the field effect limit the practical values of β to 10 or even less, which is much below the sensitivity of devices based on the edge of superconductivity. However, the latter can operate only with very low signals and cannot function at all in magnetic fields. A big advantage of semiconductor thermistors is their relative small sensitivity to magnetic fields. Ge-based bolometric detectors can operate without degradation in fields up to 8 T at temperatures of 10 mK [25]. The shift of characteristics was significant but it was possible to optimize thermometer for a given magnetic field. It follows, that a single thermometer can operate over a wide range of temperatures by "tuning" it with a variable magnetic field.

As we see, fundamental investigations of hopping conductivity are the basis for the design and technology of important devices operated at low and very low temperatures.

Chapter 6

Inter-impurity Radiative Recombination and Hopping Photoconductivity

6.1 Introduction

Similar to the hopping conductivity of equilibrium charge carriers, the transport and recombination of non-equilibrium carriers may also be determined by hopping and tunneling transitions between impurities. The "hopping recombination" mechanisms of various materials depend on the concentration and distribution of centers of localization, shape of the wave function of localized carriers, etc. In this chapter, we will discuss the inter-impurity radiative recombination (IRR) and the hopping photoconductivity (HPC) in a classical semiconducting material: lightly and heavily doped germanium in which the hopping mechanism of transport for equilibrium carriers has been well studied. Therefore, this discussion makes comparison with theoretical models more rigorous and complete.

The chapter is organized as following: we first discuss the background to the problem, general features of the inter-impurity radiative recombination model, and calculate the shape and position of the luminescence line as a function of impurity concentration and the level of excitation, and compare these results with experimental data. The next section deals with tunnel radiative recombination in heavily doped and compensated germanium. Heavy doping and compensation produce a random potential with valleys and hills of different size and amplitude. Under these conditions, carriers are localized in separated extrema of the potential. This is in contrast to inter-impurity radiative recombination, where carriers are localized on individual impurity atoms. In this case, the recombination is caused by tunneling through potential hills.

We will discuss the origin of the luminescence, its line shape and kinetics after switching off the excitation, as well as the quasi-equilibrium of excess carriers trapped in potential wells of various depths.

Finally, in the last section, we discuss hopping photoconductivity and the influence of inter-impurity and tunneling recombination on the hopping photoconductivity characteristics: dependence of photoconductivity on the level of excitation and the kinetics of conductivity.

6.2   Inter-impurity Radiative Recombination in Lightly Doped Germanium

The idea of inter-impurity radiative recombination (IRR) was proposed independently by Dobrego and Ryvkin [1, 2] in their study of hopping photoconductivity and by Williams [3] in his investigation of phosphor luminescence. Williams proposed that donors and acceptors form tightly bound pairs located in neighboring sites in the network. The main features of the IRR were developed later by Thomas and Hopfield in a series of papers devoted to luminescence of GaP samples at helium temperatures [4, 5]. IRR was studied systematically in lightly doped germanium in Refs. [6–10] and the results of these investigations were later reviewed [11].

6.2.1   *Pair model of inter-impurity radiative recombination*

Compensation of majority impurities is an necessary condition for hopping conductivity to provides empty states even at $T = 0$. In n-type material and weak compensation ($K = N_A/N_D \ll 1$), electrons from the donor are transferred to the acceptor. As a result, at low temperatures, all acceptors are charged negatively and an equal number of donors are charged positively (Fig. 6.1). Because of the Coulomb force, these charged impurities are usually neighbors among the randomly distributed impurities.

After illumination of the sample with light quanta having energy above the forbidden gap of the semiconductor, E_g, non-equilibrium electrons and holes appear (transition 1 in Fig. 6.1). After thermal relaxation within the bands, the electrons and holes are trapped by positively charged donors and by negatively charged acceptors, respectively (transitions 2 in Fig. 6.1). This makes both impurities temporarily neutral. After some lifetime τ, the non-equilibrium electrons and holes annihilate by a tunneling transition between these impurities. The rate of this transition is proportional to the overlap integral of the wave functions of the localized electrons and holes. As a result, τ increases exponentially with the distance r between

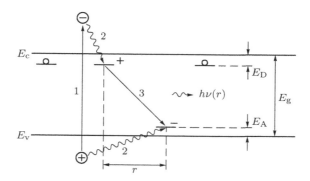

Fig. 6.1   Schematic representation of inter-impurity radiative recombination. 1 — excitation, 2 — thermal relaxation and trapping of non-equilibrium carriers, 3 — radiative recombination.

the neighboring donors and acceptors: $\tau(r) \propto \exp(2r/a)$. If this transition is radiative (IRR), the energy of the emitted light quantum $h\nu(r)$ is increased by the term $\Delta(r)$, which reflects the Coulomb energy between the charged donor and acceptor which appeared again after recombination of non-equilibrium charge carriers:

$$h\nu(r) = E_0 + \Delta(r), \qquad E_0 = E_g - E_D - E_A \pm E_{ph}. \qquad (6.1)$$

Here, E_g is the forbidden gap, E_D and E_A are the donor and acceptor ionization energies (Fig. 6.1) and E_{ph} is the energy of the emitted $(-)$ or absorbed $(+)$ phonon.

The density of charged donor–acceptor pairs is determined by minority impurities (acceptors, N_A) while the average distance R within the pair is determined by concentration of the majority impurities, $R \approx N_D^{-1/3}$. Therefore, in the case of weak compensation $(K \ll 1)$, R is much smaller than the distance between pairs, $N_A^{-1/3}$, which allows one to neglect the influence of other charges on $\Delta(r)$. Then, $\Delta(r)$ is simply

$$\Delta(r) = \frac{e^2}{\kappa r}, \qquad (6.2)$$

where κ is the dielectric constant of the host semiconductor.

Note that $\Delta(r)$ is the only parameter in this *pair model*. It contains all the specific features of IRR. Expression (6.2) shows that $\Delta(r)$ and hence $h\nu(r)$ increase with decreasing r. In other words, more closely situated pairs emit more energetic photons and for such pairs the transition time is shorter. That is sufficient to reconstruct the following specific features of IRR.

(i) The line width of IRR is defined by the r distribution. The entire line is situated at energies higher than E_0 because $\Delta(r) > 0$.

(ii) If the carrier cross-section of the trapping of non-equilibrium carriers does not depend on r, then all pairs at low excitation intensity will make an equal contribution to recombination. The luminescence peak will thus correspond to the mean distance between impurities, R. That is, $h\nu_{max} = E_0 + e^2/(\kappa R)$. However, when the excitation level is increased, the longer lifetime of the more separated pairs leads to their saturation. Since these pairs emit photons of lower energy, the luminescence maximum will shift toward higher energies.

(iii) Since the lifetime depends exponentially on the electron–hole separation, the luminescence decay will have a long tail accompanied by a shift of the luminescence peak towards lower energies.

These three characteristic features enable us to identify IRR among other recombination mechanisms. We now turn to the experimental data of the low-temperature luminescence of lightly doped and compensated Ge.

Studies have been carried out for a series of neutron-transmutation-doped p-Ge samples doped with Ga acceptors and As donors and compensation $K = N_D/N_A = 0.4$. For these samples, the model can be constructed quite rigorously because all the parameters of expression (1) are known, namely, $E_g = 746$ meV at 4.2 K [12], $E_{Ga} = 11$ meV [13], and $E_{As} = 14$ meV [14]. Germanium is an indirect band gap semiconductor. This means that the top of the valence band and the bottom of the conduction band do not occur at the same point in the Brillouin zone. In Ge, the former is at the zone center and the latter is at the zone edge in the $\langle 111 \rangle$ directions. Optical transitions are therefore indirect and momentum conservation requires that they occur either with the emission or absorption of a phonon which matches the momentum difference of the points in the Brillouin zone at which the band extrema occur. There are four such phonon modes: transverse acoustic (TA), longitudinal acoustic (LA), longitudinal optic (LO) and transverse optic (TO) phonons having energies 7.8, 27.6, 30.6, and 36.15 meV respectively [15].

For electrons localized on the impurities in indirect semiconductors, such as Ge, non-phonon recombination with holes is possible, in principle, because the impurity nucleus can play the role of the third particle needed for momentum conservation. The As impurity in Ge is characterized by the enhanced probability for a localized electron to be in the impurity nucleus $|\Psi(0)|^2$ due to a non-Coulomb potential in the central cell. This produces an intensive non-phonon (NP) line in the luminescence due to electrons localized on the As impurity in Ge.

Figure 6.2 shows the photoluminescence of a NTD-Ge sample with $N_{Ga} = 4 \times 10^{15}$ cm^{-3}. The two luminescence lines lie below the absorption edge at 746 meV. As the first approximation, we assume that $\Delta = 0$. Then the highest maximum in the spectrum lies close to the calculated position of NP line of IRR, whereas the second maximum is near the line associated with the emission of an LA phonon. This justifies the assignment of the principal maximum to the non-phonon transition from which the Coulomb interaction energy Δ of the impurities can be determined.

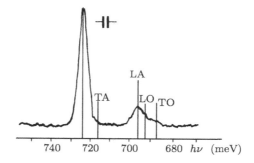

Fig. 6.2 IRR spectrum of As–Ga pairs in Ge. The main peak corresponds to non-phonon (NP) transition. The vertical lines indicate the calculated locations of lines associated with the emission of various phonons.

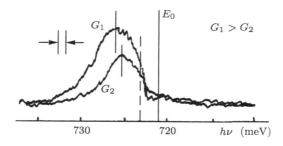

Fig. 6.3 NP line with enhanced resolution for two intensities of excitation $G_1 > G_2$. The solid vertical line refers to a NP transition between non-interacting infinitely distant pairs. The dashed line refers to a transition between impurities located at the average distance from each other.

Figure 6.3 shows the NP line measured at higher resolution at two levels of excitation. The solid vertical line corresponds to the E_0 position, and the dashed line refers to the transition between impurities located at an average distance R from each other. It can be seen that the line is shifted from E_0 toward higher energies, in agreement with point (i). One also finds that an increase in the excitation level G shifts the luminescence maximum toward higher energies, as predicted by point (ii).

Figure 6.4 shows the same NP line measured after different delay times, which were changed by varying the modulation frequency of the excitation. The lower the frequency, the longer is the delay time. One finds that an increase in delay time shifts the luminescence maximum toward lower energies, as expected from point (iii).

The experimental data thus confirm that the low temperature luminescence in lightly doped and compensated germanium is due to donor–acceptor recombination.

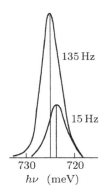

135 Hz

15 Hz

730   720

$h\nu$ (meV)

Fig. 6.4 NP radiation line at two different delay times. The higher frequency corresponds to a smaller delay time. The sample is the same as in Fig. 6.2.

6.2.2 *The peak position and the shape of inter-impurity radiative recombination line*

The pair model enables us to derive an analytic expression for the IRR line as a function of the excitation level and concentrations of the majority and compensating impurities. It also permits us to determine the dependence of the line shape on delay time. The general idea of the calculation is to divide all donor–acceptor pairs in groups according to their separation. The kinetic equation for the i-th group can be written as following:

$$\frac{dn_i}{dt} = G(N_i - n_i) - \frac{n_i}{\tau_i}, \tag{6.3}$$

where G is the generation rate per localized state, $G = \eta \alpha I / N$ is proportional to the absorption coefficient α, the quantum efficiency η and the photon flux I, $N = \sum N_i$ is the total number of localized states in pairs, n_i is the electron concentration in the i-th group.

Equation (6.3) is based on the following assumptions:

(1) In equilibrium, all pairs are unoccupied. Otherwise, n_i should be replaced by Δn_i. Indeed, when the pairs are randomly distributed, the normal situation corresponds to unoccupied donors and acceptors (1-complex). However, there are some acceptors in the neutral state (0-complex) and some donors surrounded by two charged acceptors (2-complex). Since the concentration of 0- and 2-complexes is only about 3 % of the total [16], it can be neglected.

(2) The cross-sections of excess carriers are equal for all pairs. In fact, for nearby pairs, the negatively charged acceptor pushes the electron off the donor, thus reducing the cross-section relative to distant pairs. However, this difference between cross-sections of nearby and distant pairs is much less than that between the recombination probabilities and thus can be neglected. This effect can be of significance only when the electron is trapped first because the trapped hole would neutralize the acceptor charge.

(3) The recombination velocity on a donor is proportional both to the electron concentration n_i and the probability of inter-impurity transition $\omega_i = \tau^{-1}$ and also to the probability Θ of finding a hole on the nearest acceptor. Therefore, the second term on the right side of Eq. (6.3) should be multiplied by Θ. But in fact, Θ is close to unity. At sufficiently high temperatures, when the holes are distributed randomly on the acceptors, Θ can be expressed as follows:

$$\Theta = \frac{N_A - N_D + \Delta p}{N_A} > \frac{N_A - N_D}{N_A} = 1 - K, \qquad (6.4)$$

where $K = N_D/N_A$ is the compensation ratio in p-type semiconductors ($N_A > N_D$). As the temperature decreases, most charged acceptors will be located near donors, which strengthens the inequality (6.4), implying that $\Theta = 1$ at low compensation.

In the steady state, Eq. (6.3) leads to

$$n_i^{ss} = \frac{N_i G \tau_i}{1 + G \tau_i} \qquad (6.5)$$

or, in differential form,

$$dn(r) = \frac{N(r) dr}{1 + 1/(G\tau(r))}. \qquad (6.6)$$

The intensity of recombination is then

$$S(r)dr = \frac{dn(r)}{\tau(r)}. \tag{6.7}$$

This yields

$$S(r) = \frac{N(r)}{\tau(r) + 1/G}. \tag{6.8}$$

According to Reiss [17], the distribution function $N(r)$ in the pair model is given by

$$N(r) = 3N_{ci}\frac{r^2}{R^3}\exp\left(-\frac{r^3}{R^3}\right), \tag{6.9}$$

where $R = [(4\pi/3)N_0]^{-1/3}$ is the mean distance between impurity atoms, determined by the concentration of majority impurities N_0, and N_{ci} is the concentration of the compensating minority impurities which is equal to the pair concentration.

One may evaluate $\tau(r)$ using the expression proposed by Miller and Abrahams [18] for the overlap integral of two hydrogen-like wave functions of shallow impurities with nearly equal Bohr radii a:

$$\tau(r) = \tau_0\left(\frac{2r}{a}\right)^{-7/2}\exp\left(\frac{2r}{a}\right). \tag{6.10}$$

Substituting Eqs. (6.9) and (6.10) into Eq. (6.8) and introducing dimensionless parameters $x = r/R$ and $\beta = 2r/a$, one obtains

$$S(r) = \frac{3N_{ci}}{\tau_0 R}\frac{x^2\exp(-x^3)}{(\beta x)^{-7/2}\exp(\beta x) + j^{-1}}, \tag{6.11}$$

where $j = \tau_0 G$. In Eq. (6.1), the parameter Δ depends on r and thus defines the line shape. Since $x = r/R = P/\Delta$, where $P = e^2/(\kappa R)$, we obtain the line shape:

$$S(\Delta) = \frac{3N_{ci}}{\tau_0 P}\frac{x^4\exp(-x^3)}{(\beta x)^{-7/2}\exp(\beta x) + j^{-1}}. \tag{6.12}$$

The peak position is determined from $\partial S(\Delta)/\partial x = 0$

$$4 - 3x^3 = \frac{(\beta x) - (7/2)}{1 + j^{-1}(\beta x)^{7/2}\exp(-\beta x)}. \tag{6.13}$$

The excitation intensity G or I enters in Eqs. (6.12) and (6.13) through $j = \tau_0 G = \tau_0\eta\alpha I/N_{ci}$, which we shall denote as *the effective excitation level*. When this value is very low, $j \to 0$, the peak position can be derived from $4 - 3x_0^3 = 0$ yielding $x_0 = 1.1$. This leads to the dependence $\Delta_0 =$

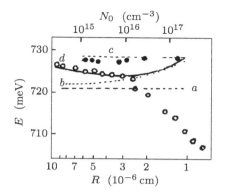

Fig. 6.5 Energy position of the IRR maximum as a function of the impurity concentration. Open circles correspond to $K = 0.4$, full circles — to $K \ll 1$. The result of calculation is shown by the full line d. Line a corresponds to undisturbed position of NP line. The Coulomb term is included in curve b. Curve c corresponds to a very high excitation level.

$0.91P \sim N_0^{1/3}$, shown in Fig. 6.5 as curve b. At high excitation, $j \to \infty$, the peak position is $\Delta_0 = 0.26e^2/(\kappa a)$ and hence does not depend on the excitation level. This case is shown in Fig. 6.5 by curve c for $a = 4$ nm.

For low and high excitations, the line shape is described by the following equations:

$$S_0(\Delta) = \frac{3N_{\mathrm{ci}}}{\tau_0 P} j x^4 \exp(-x), \qquad (6.14)$$

$$S_\infty(\Delta) = \frac{3N_{\mathrm{ci}}}{\tau_0 P} \frac{x^{1/2} \exp(-x^3)}{(\beta x)^{7/2} \exp(\beta x)}. \qquad (6.15)$$

Equation (6.15) implies that at the highest excitation level, the amplitude and line shape do not depend on the excitation level since all groups of impurity pairs are saturated. A further increase in excitation will lead to excess carriers accumulating in the valence and conduction bands and to recombination mechanisms different from IRR. The results of the line shape calculations are shown in Fig. 6.6.

For $K \ll 1$ (curve a, high level of excitation), good agreement with experiment is seen. However, for intermediate compensation (curve b), luminescence quanta with very small Δ are observed. This is due to screening of the Coulomb interaction by the redistribution of neighboring electrons, which is especially significant for intermediate compensation.

We now analyze the dependence of the peak position on impurity concentration at fixed compensation. The peak position is influenced by two factors. An increase of the main impurity concentration N_0 leads to a decrease of R, which in turn results in an increase of the Coulomb interaction and to a shift of the luminescence peak to higher energies. However, the simultaneous growth of the compensating impurity concentration N_{ci} leads

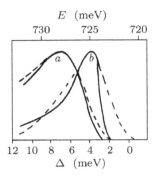

E (meV)

730   725   720

12 10 8 6 4 2   0

Δ (meV)

Fig. 6.6 Comparison between calculated (solid) and experimental (dash) line shapes. Since the effective level of excitation j is inversely proportional to K, curve a ($K \ll 1$, $N_{Ga} = 10^{16}$ cm^{-3}) corresponds to a much higher excitation level than curve b ($K = 0.4$, $N_{Ga} = 5 \times 10^{15}$ cm^{-3}).

to a decrease of the effective excitation level j at fixed illumination intensity and hence to a shift of the peak to lower energies. The second factor applies when impurity concentrations are low. At higher concentrations, when the effective excitation level become small, the first factor dominates. Hence, the dependence of Δ_0 on N at $K = $ const is non-monotonic. An example of such a calculation is shown in Fig. 6.5 (curve d). To estimate j, we used the peak position of the sample with minimal doping. Comparison with experiment shows good agreement up to $N_0 \sim 5 \times 10^{15}$ cm^{-3} (Fig. 6.5).

The shift of the luminescence peak to lower energies with increasing impurity concentration means that the main role in the recombination is played by more distant pairs. Simultaneously, the IRR line becomes narrower because the Coulomb term Δ, which determines the line width, is smaller for more distant pairs. The line narrowing is accompanied by an increase in amplitude because the integral luminescence intensity remains constant provided that non-radiative transitions are absent. Experimentally, one finds that the integral intensity is not constant (Fig. 6.7); it attains a maximum at $N_0 = 5 \times 10^{15}$ cm^{-3} and decreases at higher or lower values of N_0.

This is evidence for non-radiative recombination and can be explained in the following way. As N_0 and N_{ci} increase at low impurity concentrations, the channel for inter-impurity recombination increases. Therefore, the density of excess carriers in the conduction and valence bands decreases, which results in corresponding decrease of non-radiative recombination (through deep centers, for example). However, a further increase of impurity concentration leads to formation of impurity bands and a corresponding increase in the carrier mobility. The formation of impurity bands leads again to an increase in non-radiative recombination. Thus, $N_0 = 5 \times 10^{15}$ cm^{-3} can be regarded as the upper limit of low doping. It is worth mentioning that

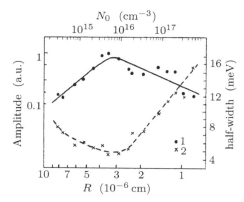

Fig. 6.7 1 — amplitude (log scale) and 2 — half-width of the IRR line as a function of impurity concentration for the same series of samples with $K = 0.4$ shown in Fig. 6.5.

at approximately the same impurity concentration ($5 \times 10^{15} \div 10^{16}$ cm^{-3}), the value of the hopping activation energy ($\varepsilon_3 \sim N^{1/3}$ for lightly doped Ge) begin to decrease with increasing N, which was also explained by the formation of an impurity band (Chapter 2, Fig. 2.5).

6.3   Tunnel Radiative Recombination of Localized Carriers in Heavily Doped and Compensated Germanium

An increase in impurity concentration eventually causes a transition from the semiconductor conductivity to the metallic state. This transition is associated with a delocalization of the equilibrium carriers. However, strong compensation causes the carriers to localize again. But now the localization does not occur on individual impurities but in the extrema of the random spatial potential fluctuations, called a potential relief, so that the recombination of excess carriers at low temperatures occurs again by tunneling.

At first sight this is a major difference, because Eq. (6.2), which determines the specific feature of IRR, is no longer valid. Briefly, this specific feature can be expressed as the relation between the luminescence photon energy and the distance between localized excess carriers under recombination: the larger the probability of tunneling, the higher the energy of the emitted photons.

One can show that a similar relation persists for tunnel radiative recombination (TRR) in heavily doped and compensated Ge (HDC-Ge). The similarities and differences between IRR and TRR mechanisms will be discussed below and we will see that the difference of the respective localization mechanisms is exhibited in the spectra and the kinetics of luminescence.

Figure 6.5 shows that an increase of N_0 above 5×10^{15} cm^{-3} leads to the rapid shift of the line peak below the "zero" energy E_0 (line a), which corresponds to the emission of a photon without the Coulomb term Δ (Fig. 6.8(a)). This shift results from the formation of a random potential relief due to fluctuations in the spatial distribution of charged impurities which is characteristic of heavily doped semiconductors. The decrease in the energy of the emitted photons is due to the localization of excess carriers in the potential valleys and hills with subsequent tunnel recombination (Fig. 6.8(b)). The dependences of the luminescence peak position on the decay time and on the excitation intensity are shown in Figs. 6.9 and 6.10 for heavily doped and compensated germanium (HDC-Ge) for $N_{Ga} = 2 \times 10^{17}$ cm^{-3}. This luminescence behavior is very similar to that observed in lightly doped Ge. This permits us to use the tunnel recombination model to explain the experimental data.

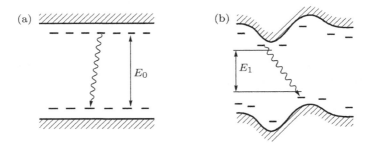

Fig. 6.8   Recombination transition schemes for low-impurity concentration (a) and for heavily doped and compensated semiconductor (b).

In a random potential relief, the mean depth of the potential well γ is related to its size r by the relation [16]:

$$\gamma(r) = \frac{e^2}{\kappa} \sqrt{Nr}. \tag{6.16}$$

Here, N is the total concentration of charged impurities. The probability for the existence of fluctuations of size r with a depth $\varepsilon \gg \gamma(r)$ decreases exponentially as $\exp\{-[\varepsilon/\gamma(r)]^2\}$. This implies that the deeper the wells, the greater is the average distance between them. Equation (6.16) leads to the same specific feature of luminescence as in IRR: the higher the energy of the emitted photons, the shorter the lifetime. The difference between IRR and tunnel radiative recombination (TRR) is that in IRR the relation

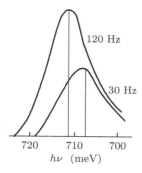

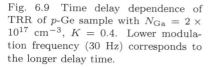

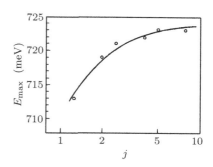

Fig. 6.9   Time delay dependence of TRR of p-Ge sample with $N_{Ga} = 2 \times 10^{17}$ cm^{-3}, $K = 0.4$. Lower modulation frequency (30 Hz) corresponds to the longer delay time.

Fig. 6.10   Shift of the luminescence peak for p-Ge sample with $N_{Ga} = 2 \times 10^{17}$ cm^{-3}, $K = 0.4$ as a function of effective excitation intensity j.

between the transition probability and the photon energy is strict, whereas in TRR it is fulfilled only on the average.

The maximal size of a random potential relief is determined by the screening radius r_s which is determined by the concentration of remaining electrons, n, i.e. by the compensation,

$$r_s = N^{1/3} n^{-2/3} = N^{-1/3}(1 - K)^{-2/3}. \tag{6.17}$$

Therefore, r_s increases with increasing K. Correspondingly, the amplitude of the potential relief ε_0 also increases:

$$\varepsilon_0 \equiv \gamma(r_s) = 2\sqrt{\pi}\alpha\frac{e^2}{\kappa}N^{1/3}(1 - K)^{-1/3}. \tag{6.18}$$

Here α is a numerical constant of order unity. Thus, an increase of compensation results in a shift of the TRR line towards low energies and a decrease of the average recombination rate. The influence of compensation on TRR was studied [19, 20] on a series of n-Ge samples initially doped with As donors at a concentration of $N_{As} = 5 \times 10^{17}$ cm^{-3} and compensated with Ga acceptors through the neutron transmutation doping technique. In this method, the degree of compensation K is a function of the irradiation dose, and K was also controlled by Hall effect measurements.

Typical luminescence spectra for various K are shown in Fig. 6.11. As K increases, the maximum of the line shifts towards lower photon energies. The luminescence intensity first increases and then decreases and the line width becomes greater. The luminescence decay is shown in Fig. 6.12. The

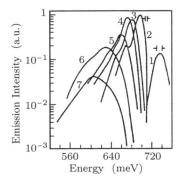

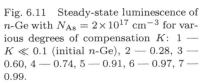

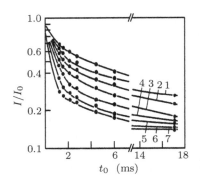

Fig. 6.11 Steady-state luminescence of n-Ge with $N_{As} = 2 \times 10^{17}$ cm^{-3} for various degrees of compensation K: 1 — $K \ll 0.1$ (initial n-Ge), 2 — 0.28, 3 — 0.60, 4 — 0.74, 5 — 0.91, 6 — 0.97, 7 — 0.99.

Fig. 6.12 Luminescence decay for several photon energies measured for sample with $K = 0.74$. $h\nu$, meV: 1 — 658, 2 — 664, 3 — 671, 4 — 674, 5 — 678, 6 — 683, 7 — 687.

greater the photon energy, the faster the decay. However, the decay is non-exponential; the decay rate decreases with time for fixed photon energy.

These features can be explained within the model of radiative recombination of non-equilibrium carriers localized in the extrema of the random potential relief.

The monotonic shift of the peak position to lower energies reflects the increase of the mean depth of the potential wells of the random potential with increasing K. Equation (6.18) shows that this shift must be proportional to $\varepsilon_0 \sim (1 - K)^{-1/3}$. Figure 6.13 shows that for $K > 0.6$, the position of the line maximum, indeed, follows $(1 - K)^{-1/3}$. The slope gives $\alpha = 1.3$.

The non-monotonic dependence of the luminescence intensity with increasing compensation can be explained by assuming that for a heavily doped sample with $K \ll 0.1$ (curve 1 in Fig. 6.11), there also exist some non-radiative recombination centers N_{nr}. Therefore, the quantum efficiency of the luminescence is relatively small. The mean distance between these non-radiative recombination centers is $R_{nr} \approx N_{nr}^{-1/3}$. As a result, the non-equilibrium charge carriers have to diffuse over this distance before being trapped by these centers and recombining without emitting light. An increase in compensation of up to $K = 0.75$ strongly suppresses the diffusion of non-equilibrium carriers. This results in an increase in intensity of radiative recombination (curves 2–4 in Fig. 6.11). However, for $K \to 1$, the maximal size of the potential wells r_s increases continuously (Eq. (6.17)).

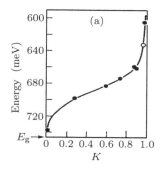
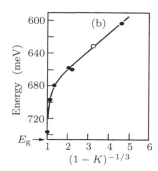

Fig. 6.13   Energy position of the line maximum as a function of K (a) and $(1 - K)^{-1/3}$ (b).

When $r_s > R_{nr}$, almost every potential well contains a non-radiative recombination center and the quantum efficiency again decreases. From the strong quenching of the luminescence at $K > 0.9$, one can determine R_{nr}, and thus estimate $N_{nr} \approx 10^{16}$ cm^{-3}.

The shift of the peak position to higher energies with increasing excitation intensity (Fig. 6.10) is explained by the fact that deep wells are characterized on average by a large size and therefore by a smaller probability of tunneling and a longer lifetime. Recombination from deep wells is accompanied by emitting light quanta having low energy. An increase of excitation intensity causes the deep wells to become saturated, and the main contribution to the luminescence is shifted to more shallow wells with shorter lifetime, similar to the case of IRR.

We now discuss the decay of luminescence in TRR. Let us assume that the recombination of non-equilibrium carriers occurs from potential wells with depth ε_0. Equation (6.16) implies that wells of such depth are mainly associated with the fluctuations having linear size $r_0 = \varepsilon_0^2/(e^2/\kappa)^2 N$. The probability is exponentially small of the appearance of a well having depth ε_0 and $r \ll r_0$. However, one should take into account that as the concentration of small size fluctuations increases, it grows as $(r/r_0)^3$. Hence, the following equation describes the concentration of potential wells with depth ε_0 and size $r \ll r_0$,

$$N(\varepsilon_0, r) = N(\varepsilon_0, r_0)(r/r_0)^3 \exp(-r/r_0). \qquad (6.19)$$

It can be seen that $N(\varepsilon_0, r)$ is a smooth function of r in the range from $0.2r_0$ to r_0. This means that for photons of fixed energy, there exist a number of characteristic decay times. In this respect, the recombination of excess carriers localized in potential wells differs substantially from

IRR where the relation between $h\nu$ and r is fixed. This explains the non-exponential decay for fixed photon energy and the spectral dependence of the decay (Fig. 6.12). With increasing excitation level, the major role in recombination is transferred to fast recombination channels because the slow channels are filled. This explains why the decay is faster for more energetic luminescence quanta. The shift of the luminescence peak toward higher energies can be explained in the same way (Fig. 6.10). This shift is less pronounced than that observed in IRR because there is now a set of decay times for every photon energy.

6.4   Radiative Recombination in Ge–Si Solid Solution

The investigation of hopping conductivity in the samples with solid solution $Ge_{1-x}Si_x$ ($x \leq 0.08$) described in Chapter 3 showed the following: the increase of the energy activation of hopping conductivity ε_3 as a function of x can be explained by theoretical calculations only under the assumption that Si atoms do not penetrate into Ge lattice separately, but in the form of clusters with $\Theta \approx 50$ atoms on average. To verify this assumption, we have measured the IRR for the same series of samples [21], because the analysis of the luminescence line can also give information about the spatial distribution of impurities, their energy positions, and the wave function of localized charge carriers.

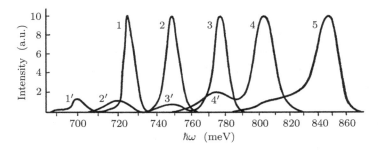

Fig. 6.14   Typical spectra IRR As–Ga for $Ge_{1-x}Si_x$ at 4.2 K; x: 1 — 0, 2 — 0.015, 3 — 0.04, 4 — 0.06, 5 — 0.08, 1′–4′ —LA lines.

The concentration of impurities was the same in all samples, $N_{Ga} = 3.5 \times 10^{15}$ cm^{-3}, $N_{As} = 1 \times 10^{15}$ cm^{-3}. The only variable parameter was the value of x. Figure 6.14 shows typical spectra of IRR As–Ga for the initial sample ($x = 0$) and for samples with $x = 0.015$, 0.04, 0.06 and 0.08 (which are the average values x^* over a sample).

It is seen that with increasing x, the position of the lines shifts to higher energies, the lines are broadened and the relation is changed between the main non-phonon (NP) line and its phonon (LA) replica. These dependences are shown in Figs. 6.15–6.17.

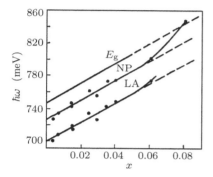

Fig. 6.15   Position of the maximum of NP and LA lines as a function of x.

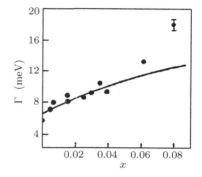

Fig. 6.16   Half-width of the NP line as a function of composition x.

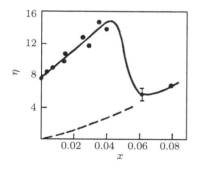

Fig. 6.17   Ratio between intensities of NP and LA lines in samples having different x.

For $x > 0.04$, remarkable deviations from monotonic dependences are observed. Therefore, we first discuss the data obtained for samples with $x \leq 0.04$.

6.4.1 *Position of the luminescence maximum*

For IRR, the expression for the energy of emitted phonon is

$$h\nu = E_{\mathrm{g}} - E_{\mathrm{D}} - E_{\mathrm{A}} \pm E_{\mathrm{ph}} + \frac{e^2}{\kappa r}. \tag{6.20}$$

Here, E_{D} and E_{A} are the energy of ionization for an isolated donor and acceptor, E_{ph} is the energy of the emitted $(-)$ or absorbed $(+)$ phonon, and the last term characterizes the energy of the Coulomb interaction between a charged donor and a charged acceptor separated by a distance r.

Figure 6.15 shows the position of the main non-phonon line (NP), and its LA-replica as a function of x, together with the dependence of the forbidden gap $E_{\mathrm{g}}(x)$ in solid solution Ge–Si [22]. All curves are parallel, which means that all changes are determined by an increase of the forbidden gap E_{g}. Within the accuracy of the measurements, the values of E_{D}, E_{A} and E_{ph} are not changed within the interval $x \leq 0.04$.

6.4.2 *Half-width of the luminescence line*

At $x = 0$, the width of the IRR line is determined by the scatter of the distance r within the pairs of donors and acceptors: more separated pairs emit light quanta with smaller energy. In a solid solution, an additional scatter appears caused by the local fluctuations of the composition δx. This in turn, leads to fluctuations of the value of the forbidden gap $E_{\mathrm{g}} = E_{\mathrm{c}} - E_{\mathrm{v}}$, or, to be more exact, to fluctuations in the energy of the bottom of the conduction band δE_{c} and the top of the valence band δE_{v}. Obviously, the smaller the volume, the greater the fluctuations. The maximal effect corresponds to fluctuations on the scale of the Bohr radius a_{B}, because fluctuations of smaller size will be averaged by the wave function. A specific feature of the IRR is that there are tunneling transitions of carriers between different regions of a crystal. In one region a shallow donor is located, and in the other region, a shallow acceptor, and each follows the fluctuations of "its own" band edge: the bottom of the conduction band δE_{c} or the top of the valence band δE_{v}. Assuming that these values fluctuate independently, one can write the expression for the luminescence line half-width

$$\Gamma = \sqrt{\Gamma_0^2 + \delta E_{\mathrm{c}}^2 + \delta E_{\mathrm{v}}^2}. \tag{6.21}$$

Here, Γ_0 is an initial half-width (at $x = 0$),

$$\delta E_{\mathrm{c,v}} = \frac{\alpha_{\mathrm{c,v}}}{2\sqrt{\pi}} \sqrt{\frac{x}{N a_{\mathrm{e,h}}^3}}, \tag{6.22}$$

where $\alpha_{c,v} = dE_{c,v}/dx$ at $x = x^*$, $N = 4.45 \times 10^{22}$ cm^{-3} is the concentration of Ge atoms in the lattice, $a_{e,h}$ is the Bohr radius of an electron on a donor (e) or a hole on an acceptor (h). This equation was already given in Chapter 3 under the assumption that the Si atoms are distributed separately in Ge lattice. If one assume that Si atoms form the clusters of Θ atoms, the amplitude of fluctuations will be enhanced by the factor $\sqrt{\Theta}$:

$$\delta E_{c,v} = \frac{\alpha_{c,v}}{2\sqrt{\pi}} \sqrt{\frac{\Theta x}{N a_{e,h}^3}}. \tag{6.23}$$

We need to know the values of α_c and α_v. It is known that at small x, the forbidden gap of Ge$_{1-x}$Si$_x$ increases with coefficient $\alpha_g = 1.2$ eV [22] and $\alpha_g = \alpha_c + \alpha_v$. It is also known that in the heterojunction Ge–Si, the shift of the valence band is 0.2 eV [23]. If this change is uniform for all x, $\alpha_v \approx 0.2$ eV, and therefore $\alpha_c \approx 1$ eV. This estimate implies that $(\delta E_c)^2 \gg (\delta E_v)^2$. Therefore, the dependence $\Gamma(x)$ can be written as

$$\Gamma(x) = \sqrt{\Gamma_0^2 + Ax}, \qquad A = \frac{\alpha_c}{2\sqrt{\pi}} \sqrt{\frac{\Theta}{N a_e^3}}. \tag{6.24}$$

Inserting the values of α_c, N and $a_e = 4$ nm into Eq. (6.24) gives $A = 5.2\Theta^{1/2}$ meV. The best fit to the experimental curve shown in Fig. 6.16 corresponds to $\Gamma_0 = 6.5$ meV, $A = 38$ meV. Comparison with theoretical value of A allows one to estimate the average number of Si atoms in clusters: $\Theta \approx 50$. This value is in agreement with the data obtained from of the hopping conductivity data in the same series of samples (Chapter 3).

6.4.3 Relation between non-phonon and phonon bands

Germanium is an "indirect" semiconductor and therefore the optical transitions between non-equilibrium electrons and holes are accompanied by participation of a "third" particle, usually, a longitudinal acoustic (LA) phonon. In Ge$_{1-x}$Si$_x$ solid solution, a non-phonon (NP) optical transition may appear due to scattering by the short-range potential of Si atoms. It was shown that the ratio η between NP and LA lines in "interband" or "exciton" luminescence, is zero at $x = 0$, increases with x, and reaches the value of 2.7 at $x = 0.045$ [24]. This dependence is shown in Fig. 6.17 as a dotted line.

In our samples, however, due to the non-Coulomb potential in the central cell of As donors, the probability of NP transition is very high even without Si, with the ratio $\eta(0) = 7$. In this case, it seems that the additional probability for NP transitions, connected with Si atoms in Ge–Si samples would increase η very slightly, but experiment shows that this is not so: the ratio $\eta(x)$ increases much faster than for the exciton luminescence (Fig. 6.17). This effect can be explained by assuming the formation of As–Si complexes with As and Si atoms in nearest-neighbor lattice sites which are characterized by a very sharp potential of the central cell. This potential provides the enhanced probability of NP transitions. At small x, the probability of such complexes increases linearly with x, which explains the linear dependence of $\eta(x)$.

6.4.4 *Radiative recombination in* $Ge_{1-x}Si_x$ *at* $x > 0.04$

Only two samples were investigated [21] with $x = 0.06$ and $x = 0.08$, but the changes in the luminescence spectra in these samples were very significant:

(a) The position of the line maximum shifts rapidly toward high energies, which corresponds to the "interband" recombination instead of IRR — "interimpurity" recombination (Fig. 6.15).

(b) The half-width of the NP line increases more rapidly (Fig. 6.16) and the shape of the lines is also changed (Fig. 6.14); they become more extended toward low energies, which is characteristic of tunnel radiative recombination (TRR) of carriers localized in the extrema of the random potential relief.

(c) IRR is characterized by a strong dependence of the decay kinetics (lifetime τ) on the energy $h\nu$ of the emitted quanta (curve 1 in Fig. 6.18). For TRR, this dependence is much weaker, because there is a set of decay times for every photon energy due to localization of charge carriers in potential wells of different depth and size. The data show that for $x > 0.04$, the dependence of $\tau(h\nu)$ becomes barely visible (curves 6 and 7 in Fig. 6.18).

(d) The value of $\eta(x)$ sharply decreases and become comparable with the values which would be observed at the "interband" or "exciton" luminescence without the participation of an As impurity.

We believe that all these features can be explained by the assumption that for $x > 0.04$, non-equilibrium electrons are localized in potential wells caused by fluctuations in the spatial distribution of Si atoms, rather than by individual As impurities. These wells have to be deep enough, since localization in the area of radius r leads to an additional kinetic energy for electrons $E_{\mathrm{kin}} \sim \hbar^2/2m^*r^2$. Localization requires that the depth of the well $E_{\mathrm{pot}} > E_{\mathrm{kin}}$. The value of E_{pot} could be estimated from Eq. (6.23). Let us take $r = a$, where $a = 4$ nm is the Bohr radius of an electron localized on an As impurity in Ge. This gives $E_{\mathrm{kin}} = 8$ meV. To estimate E_{pot}, we take $\alpha_c = 1$ eV, $\Theta = 50$. Inserting these values into Eq. (6.23) shows that E_{pot} increases E_{kin} just at $x = 0.04$.

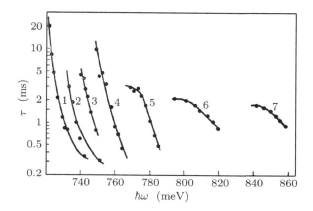

Fig. 6.18 Dependence of the effective decay time on the energy of emitted quanta for NP line for samples with different composition, x: 1 — 0, 2 — 0.007, 3 — 0.015, 4 — 0.03, 5 — 0.04, 6 — 0.06, 7 — 0.08.

This estimate supports the assumption about the change in the character of localization of non-equilibrium electrons at $x > 0.04$. On the other hand, this confirms the necessity to take into account the formation of Si clusters in the Ge lattice with the average number of atoms in a cluster being $\Theta \approx 50$.

6.5 Thermal Quenching of Luminescence

The temperature quenching of the luminescence was observed in both lightly doped and heavily doped samples [9]. Figure 6.19 shows that the quenching is accompanied by a shift of the peak position toward low energies, which is more pronounced for highly doped samples. Quenching occurs at relatively low temperatures in a narrow interval $4.2 \div 20$ K. Experiments showed that in the interval below 4.2 K no significant changes take place.

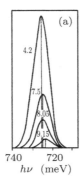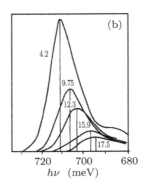

Fig. 6.19   Thermal quenching of IRR (a) and TRR (b) lines. The numbers alongside the curves indicate the temperature in K; N_{Ga}, cm^{-3}: (a) — 2×10^{15}, (b) — 2×10^{17}.

Figure 6.20 shows the thermal quenching curves for IRR (lightly doped samples) and TRR (heavily doped samples); allowance is made for the difference between the luminescence intensities at 4.2 K.

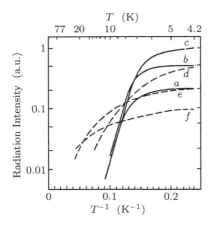

Fig. 6.20   Temperature dependence of the luminescence intensity for series of p-Ge samples having different concentrations of impurities and constant compensation $K = 0.4$. N_{Ga}, cm^{-3}: a — 7.6×10^{14}, b — 2.1×10^{15}, c — 5.0×10^{15}, d — 3.5×10^{16}, e — 2×10^{17}, f — 5.5×10^{17}.

It has already been mentioned that IRR is accompanied by non-radiative recombination. Therefore, the thermal quenching of lumines-cence can be explained as accelerated diffusion of excess carriers to the centers of non-radiative recombination. For low impurity concentration ($N_{Ga} \leq 5 \times 10^{15}$ cm^{-3}), the energy of activation of the IRR thermal quench-ing $\Delta E = 7.5 \pm 1.5$ meV remains constant. This energy can be attributed to an excited impurity level [25]. At low impurity concentrations and low temperatures, non-equilibrium carriers are captured by the ground states of shallow impurities and the probability to reach the non-radiative recombination centers is low. Increasing the temperature leads to the thermal transition of carriers to excited levels which results in an increased mobility and in enhancement of non-radiative recombination.

In the region of high impurity concentration ($5 \times 10^{15} < N_{Ga} < 5 \times 10^{17}$ cm^{-3}), impurity bands are formed, which leads to a decrease of the energy of ionization of impurities, the merging of the excited and then ground states of the shallow impurities with the tails of the main bands. The mobility of non-equilibrium carriers in the ground state increases. Therefore, the non-radiative recombination may take place even at low temperatures. As a result, the intensity of luminescence at 4.2 K decreases, the thermal quenching slows down, and there is no definite activation energy of quenching in the investigated range of temperatures.

For these samples, the increase of temperature also leads to a shift of the luminescence peak towards smaller photon energies (Fig. 6.19(b)). For heavily doped and compensated (HDC) samples, non-equilibrium carriers are localized in the extrema of the random potential relief, as described in Sec. 6.3. In our HDC p-Ge samples with $K = 40$ %, the acceptor impurity band is approximately half-filled, but the donor impurity band is empty at equilibrium. For low temperatures and weak excitation, the non-equilibrium electrons are trapped by a relatively deep potential well nearest the point of generation of this electron. Such an electron cannot be transferred to a deeper state even if that state is empty. In this case, the energy spectrum of non-equilibrium electrons does not correspond to the density-of-states function of the donor impurity band. When the temperature increases, electron transitions to the deepest states become possible and this shifts the maximum of the luminescence line.

Summarizing these sections, all specific features of the luminescence of lightly doped germanium can be explained by the model of an inter-impurity radiative recombination (IRR) of non-equilibrium charge carriers localized on donor and acceptor impurities. For heavily doped and compensated germanium, the explanation is based on model of the tunnel radiative recombinaton (TRR) of excess carriers localized in the extrema of the random potential relief. Similar dependences have been observed in the luminescence of other materials such as highly doped and compensated GaAs [26] and glassy As$_2$Se$_3$ [27], which shows that the proposed model is quite general.

6.6   Hopping Photoconductivity

Photoconductivity, meaning the change of conductivity caused by illumination, is a very wide field of semiconductors physics which covers numerous problems concerning the generation, transport and recombination of excess carriers. The transport of excess carriers can differ from that of "dark" carriers. For example, the dark conductivity may proceed by hopping among localized states whereas the photoconductivity proceeds in extended band states because the excess carriers are excited to these bands. We will use the term "hopping photoconductivity" to stress the hopping character of excess carrier transport in our samples.

It is well known that photoconductivity (PC) may be caused by a light-induced change in the concentration of excess carriers (n-PC) or in their mobility (μ-PC). Of course, the combination of these two processes is also possible. Without going into detail, μ-PC tends to appear with photon energies much less than the forbidden gap. In the opposite case, the appearance of n-PC is more probable. We here discuss the results which concern only n-PC.

6.6.1   *Negative Hopping Photoconductivity*

It should be mentioned that hopping photoconductivity (HPC) leads to a complicated relation between $\Delta n/n$ and $\Delta\sigma/\sigma$ because the illumination produces "decompensation" which may lead both to positive and negative HPC. The latter is observed in the case of small impurity concentration N, small degree of compensation $K < 0.5$ and at relatively weak illumination. The first observation of negative HPC was reported by Dobrego and Ryvkin [1]. Figure 6.21 shows the influence of illumination on hopping conductivity for three samples of n- and p-Ge observed by Davis [28]. At low intensity of illumination, the conductivity first decreases, but with further increase of illumination, the negative effect is replaced by positive photoconductivity. Negative HPC is large for the sample with small N. With increase of N, negative HPC decreases and finally disappears.

Negative HPC can be explained in the following way. For low compensation ($K < 0.5$), the value of hopping conductivity is determined by the density of the empty places. After excitation, the non-equilibrium carriers are trapped by the empty places (Fig. 6.1). This leads to a decrease in the number of empty places and a corresponding decrease of hopping conductivity. The photo-excited charge carriers which remain in the conduction (or valence) band have high mobility. Therefore, they can mask completely the hopping conduction process. As a result, increasing the intensity of illumination always leads to positive photoconductivity. An increase of N leads to the formation of an impurity band, where the existence of an empty place is not as crucial as for lightly doped samples. At the same intermediate concentrations, the energy of activation of hopping conductivity ε_3 starts to decrease (see Fig. 2.5), and the position of the maximum IRR line shifts rapidly to lower energies (see Fig. 6.5). All these phenomena are determined by the formation of an impurity band. Finally, at relatively high N, the negative effect disappears and the conductivity increases linearly with an increase of illumination (Fig. 6.21).

There is an additional aspect of negative HPC: the decrease in conductivity (or increase of resistance) under light illumination is strongly enhanced by a decrease of temperature. In lightly doped samples, as the compensation K is increased, the hopping energy of activation ε_3 passes through a minimum value at $K = 0.4$–0.5 (Fig. 6.22) [29]. If the initial

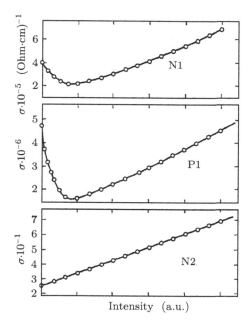

Fig. 6.21   Hopping photoconductivity for three samples n- and p-Ge as a function of light intensity. Sample N1: $N_{\text{Sb}} = 1.8 \times 10^{16}$ cm^{-3}; sample P1: $N_{\text{Ga}} = 3.7 \times 10^{15}$ cm^{-3}; sample N2: $N_{\text{Sb}} = 7 \times 10^{16}$ cm^{-3} [28].

compensation is small, the negative photoconductivity has to be accompanied by an increase of ε_3 due to decompensation (Fig. 6.23).

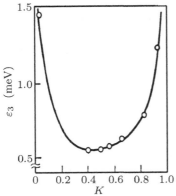

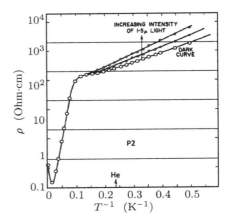

Fig. 6.22 Dependence of energy activation of hopping conductivity ε_3 on the degree of compensation K for lightly doped Ge [29].

Fig. 6.23 Temperature dependence of the conductivity of sample P2 ($N_{\mathrm{In}} = 1 \times 10^{16}$ cm^{-3}, $K = 0.4$) in dark and under different intensities of illumination [28].

6.6.2 *Positive Hopping Photoconductivity*

Hopping photoconductivity (HPC) may lead to a complicated relation between the concentration of photo-excited carriers and the corresponding change in conductivity. But in extreme cases, $K \to 1$ or $K \to 0$, the relation between $\Delta n/n$ and $\Delta\sigma/\sigma$ is close to linear. We here deal only with this situation.

We now turn to the experimental results and their interpretation. Figures 6.24–6.27 show dependencies of HPC on the excitation intensity, decay time after excitation and on the energy of illuminating photons. In principle, tunnel recombination is the same as IRR: as the time increases, the excess carriers recombine from the potential wells of gradually increasing size. The characteristic time τ grows exponentially with the size of the potential well. Hence, by analogy with IRR, one can define the parameter $r_0(G)$, where G is the generation rate (see Eq. (6.3)), for which $G\tau_0(r_0) \approx 1$ and consider that nearly all wells having $r \gg r_0$ are occupied by excess carriers, while nearly all wells having $r \ll r_0$ are empty. As G increases, r_0

decreases, and the main role is played by fluctuations of smaller size. Hence,

$$N(G) \approx \int_{r_0(G)}^{\infty} N(r)dr, \tag{6.25}$$

where $N(r)$ is the size distribution of the potential wells.

Let us assume a very weak excitation, so that $r_0 \gg r_s$, where r_s is the screening radius which defines the scale of the amplitude of the potential relief. The concentration of large-scale fluctuations is $N(r) \sim \exp(-r/r_s)$. Inserting into Eq. (6.25) gives

$$\frac{n(G)}{N} \approx r_s \exp\left(\frac{r_0}{r_s}\right). \tag{6.26}$$

To find this value of r_0 we use Eq. (6.5), where $N(r)$ is now the concentration of potential wells of given size, and the lifetime is

$$\tau(r) = \tau_0 \exp\left(\frac{2r}{a}\right). \tag{6.27}$$

The definition of $r_0(G)$ then gives

$$r_0(G) = \frac{a}{2}\ln[(G\tau_0)^{-1}] \tag{6.28}$$

Hence,

$$\frac{n(G)}{N} \approx \exp\left[\frac{a}{2r_s}\ln(G\tau_0)\right] = (G\tau_0)^{a/(2r_s)}. \tag{6.29}$$

Since $r_s \gg a$, the dependence of the photoconductivity on intensity is strongly sublinear. An example of such dependence measured at 4.2 K is shown in Fig. 6.24. The intensity dependence is sublinear with slope $a/2r_s = 0.24$ of the double logarithmic plot. This gives $2r_s/a \approx 4$.

We now discuss photoconductivity decay. The tunnel recombination model predicts non-exponential photoconductivity decay after switching off the illumination. According to this model, the decay time

$$\tau(t) = -\frac{\sigma(t)}{d\sigma(t)/dt} \tag{6.30}$$

should grow with time. With increasing time, the recombination will pass from close pairs to distant pairs. This situation is similar to that existing in steady slate HPC when the excitation intensity is decreased. Hence, the HPC decay law should be related to the intensity dependence of the photoconductivity and should depend mainly on the relative value $\Delta\sigma/\sigma$. Therefore, the decay curves measured at different intensities should coincide

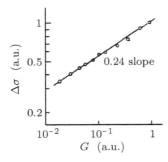

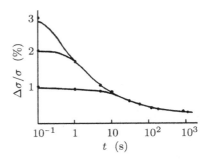

Fig. 6.24 Dependence of hopping photoconductivity of HDC-Ge on the
intensity of illumination on double-
logarithmic scale.

Fig. 6.25 Hopping photoconductivity
decay of the same sample of HDC-Ge
for different excitation levels.

when their zero point is appropriately shifted. The prediction has been
verified experimentally, as shown in Fig. 6.25.

In order to evaluate $n(t)$, let us assume that at $t = 0$, the photoconductivity has reached its steady state value which is defined by Eq. (6.29).
Then all pairs with $r > r_0$ are occupied, where r_0 is defined by Eq. (6.28).
After switching off the excitation, r_0 increases. It is clear that the majority
of pairs, which are characterized by the distance r, will be emptied during
time $\tau(r)$. Using Eq. (6.28) and the initial condition ($r = r_0$ at $t = 0$), one
can evaluate $r(t)$

$$r(t) = \frac{a}{2} \ln\left(\frac{t + G^{-1}}{\tau_0}\right). \tag{6.31}$$

Inserting Eq. (6.31) into $n(t) = N \exp[-r(t)/r_s]$ yields the following equation for the decay curve

$$\frac{n(t)}{N} = \left(\frac{\tau_0}{1 + G^{-1}}\right)^{a/(2r_s)}. \tag{6.32}$$

It follows from Eq. (6.32) that the decay is a complicated hyperbola and
that the effective decay time is

$$\tau(t) = \frac{2r_s}{a}(t + G^{-1}). \tag{6.33}$$

In accordance with Eq. (6.33), $\tau(t)$ increases linearly with slope $2r_s/a$.
Figure 6.26 shows the $\tau(t)$ dependence obtained from the $\sigma(t)$ curve of
Fig. 6.25. The slope is 3.7, which agrees well with the value $2r_s/a = 4$ estimated from the dependence of photoconductivity on the excitation
intensity (see Fig. 6.24 and Eq. (6.29)).

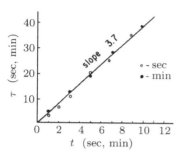
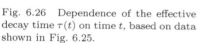
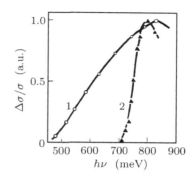

Fig. 6.26 Dependence of the effective decay time $\tau(t)$ on time t, based on data shown in Fig. 6.25.

Fig. 6.27 Spectral dependence of photoconductivity in HDC-Ge (curve 1) and in lightly doped Ge (curve 2).

We conclude this chapter by showing in Fig. 6.27 the spectral dependence of steady state photoconductivity of heavily doped and compensated (HDC) Ge in comparison with that of lightly doped Ge. In HDC-Ge (curve 1), one finds a long tail towards small photon energies which reflects the existence of large potential fluctuations.

All major features of the hopping photoconductivity thus can be explained within the model of inter-impurity and tunnel recombination of excess carriers localized on impurities or in extrema of the random potential relief.

Afterword

The reader of this book has probably noticed that almost all the references relate to work done many years ago. The question may therefore arise: why is it necessary to "stir up the past"? I think this should be done for several reasons. First of all, hopping conductivity is not just an "exotic" mechanism of electron transport observed in doped semiconductors at low temperatures. Quite the contrary. Hopping conduction is even more prevalent in nature at normal temperatures than the "classical" conductivity of free electrons. Currently, as the technology permits the synthesis of many new materials, people are beginning to study electrical signals, photoconductivity and luminescence in organic materials and biological objects, where the mechanism of electron transport is hopping. Therefore, young researchers who are studying these new materials will now and then observe phenomena that are similar to those that have already been investigated in the past on classical materials, such as doped germanium and silicon. However, they may not be aware of studies that have been carried out years ago and therefore do not refer to them. As a result, they often have to "reinvent the wheel".

The sources of information about what has been done in a given field of knowledge are usually books and reviews. The articles that comprise the content of this book (with few exceptions) are not covered in the existing books and reviews devoted to hopping conduction, partly due to the fact that they appeared after the publication of these books. An additional reason is that books and reviews are usually written by theorists who do not like to discuss problems that do not have a rigorous theoretical explanation. Since I am an experimentalist, I have no such fears. On the contrary. I believe that the publication of a book with well-established experimental facts, but yet not having a rigorous theoretical explanation, is useful because it can once again attract the interest of theorists. This includes the universal hopping prefactor and a model of "phononless" hopping conductivity, negative hopping magnetoresistance, the influence of spin polarization on two-dimensional hopping conductivity and others.

These phenomena are described and discussed here, and when similar effect will be observed by somebody else in a new material, the reader of this book will be aware of its earlier manifestation and what was the basis on which the explanatory model was built. If this will help in his or her work, one can say that this book has fulfilled its purpose.

Bibliography

Chapter 1

[1] Mott, N. F. and Kaveh, M. (1985). *Advan. Phys.* **34**, p. 329.
[2] Kravchenko, S. V. and Sarachik, M. P. (2004). *Rep. Prog. Phys.* **67**, p. 1.
[3] Mott, N. F. (1990). *Metal-Insulator Transitions*, 2nd edn. (Tailor & Francis).
[4] Altshuler, B. I. and Aronov, A. G. (1985). In: A. L. Efros, M. Pollak (eds.), *Electron-electron interaction in disordered conductors.*
[5] Abrahams, E., Anderson, P. W., Licciardello, D. C., and Ramakrishnan, T. W. (1979). *Phys. Rev. Lett.* **42**, p. 673.
[6] Lee, P. A. and Ramakrishnan, T. V. (1985). *Rev. Mod. Phys.* **57**, p. 387.
[7] Castellani, C., Di Castro, C., Lee, P. A., and Ma, M. (1984). *Phys. Rev. B* **30**, p. 527.
[8] Finkelstein, A. M. (1984). *Sov. Phys. JETP.* **59**, p. 212.
[9] Castellani, C., Kotliar, G., and Lee, P. A. (1987). *Phys. Rev. Lett.* **59**, p. 323.
[10] Sasaki, W. (1988). In: T. Ando, H. Fukujama (eds.), *Anderson Localization.*
[11] Kramer, B., MacKinnon, A. (1993). *Rep. Prog. Phys.* **56**, p. 1469.
[12] Hertel, G., Bishop, D. J., Spencer, E. G., Rowell, J. M., and Dynes, R. C. (1983). *Phys. Rev. Lett.* **50**, p. 743.
[13] McMillan, W. L. and Mochel, J. (1981). *Phys. Rev. Lett.* **46**, p. 556.
[14] Thomas, G. A., Ootuka, Y., Katsumoto, S., Kobayashi, S., and Sasaki, W. (1982). *Phys. Rev. B* **25**, p. 4288.
[15] Ootuka, Y., Matsuoka, H., and Kobayashi, S. (1988). In: T. Ando, H. Fukujama (eds.), *Anderson Localization.*
[16] Hirsh, M. J., Thomanschefsky, U., and Holcomb, D. F. (1988). *Phys. Rev. B* **37**, p. 8257.
[17] Dai, P., Zhang, Y., and Sarachik, M. P. (1991). *Phys. Rev. Lett.* **67**, p. 136.
[18] Ionov, A. N., Lea, M. J. and Rentzsch, R. (1991). *JETP Lett.* **54**, p. 25.
[19] Newman, P. F. and Holcomb, D. F. (1983). *Phys. Rev. Lett.* **51**, p. 2144.
[20] Long, A. P. and Pepper, M. (1984). *J. Phys. C* **17**, p. L425.
[21] Koon, D. W. and Castner, T. G., (1988). *Phys. Rev. Lett.* **60**, p. 1755.
[22] Dai, P., Zhang, Y., and Sarachik, M. P. (1991). *Phys. Rev. Lett.* **66**, p. 1914; (1992) *Phys. Rev. B* **45**, p. 3984.

[23] Paalanen, M. A., Rosenbaum, T. F., Thomas, G. A., and Bhatt, R. N. (1982). *Phys. Rev. Lett.* **48**, p. 1284.

[24] Stupp, H., Hornung, M., Lakner, M., Madel, O., and Lhneysen, H. V. (1993). *Phys. Rev. Lett.* **71**, p. 2634.

[25] Belitz, D. and Kirkpatrick, T. R. (1995). *Phys. Rev.* B **52**, p. 13922.

[26] Shlimak, I., Kaveh, M., Ussyshkin, R., Ginodman, V., and Resnick, L. (1996). *Phys. Rev. Lett.* **77**, p. 1103.

[27] Imry, Y. and Gefen, Y.(1984). *Phil. Mag.* B **50**, p. 203.

[28] Finkelstein, A. M. (1983). *Sov. Phys. JETP,* **57**, p. 97.

[29] Westervelt, R. M., Burns, M. J., Hopkins, P. F., Rimberg, A. J., and Thomas, G. A. (1988). In: T. Ando and H. Fukuyama (eds.), *Anderson Localization.*

[30] Malliepaard, M. C., Peper, M., Newbury, R., and Hill, G. (1988). *Phys. Rev. Lett.* **61**, p. 369.

[31] Shlimak, I., Kaveh, M., Ussyshkin, R., Ginodman, V., and Resnick, L. (1997). *Phys. Rev.* B **55**, p. 1303.

[32] Shlimak, I., Kaveh, M., Ussyshkin, V. R., Ginodman V., and Resnik, L. (1996). In: M. Scheffler and R. Zimmerman (eds.),*The Physics of Semiconductors*, vol. 1 p. 161.

[33] Shlimak, I., Kaveh, M., Ussyshkin, V. R., Ginodman V., Resnik, L., and Gantmakher, V. F. (1997). *J. Phys.: Condens. Matter,* **9**, p. 9873.

[34] Lagendijk, A., Tiggeln, B. V., and Wiersma, D. S. (August 2009). *Physics Today*, p. 24.

[35] Lagendijk, A., Tiggeln, B. V., and Wiersma, D. S. (May 2012). *Physics Today*, p. 11.

Chapter 2

[1] Shklovskii, B. I. and Efros, A. L. (1984). *Electronic properties of doped semiconductors* (Springer-Verl, Berlin; Russian original: Nauka, M.).

[2] Pollak, M. and Shklovskii, B. (Eds.) (1991). *Hopping transport in solids* (North Holland).

[3] Gantmakher, V. F. (2005). *Electrons and Disorder in Solids* (Oxford Univ. Press, N.Y.; Russian original: Fizmatlit, M.).

[4] Miller, A. and Abrahams, E. (1960). *Phys. Rev.* **120**, p. 745.

[5] Shklovskii, B. I. and Shlimak, I. S. (1972). *Sov. Phys. Semicond.* **6**, p. 104.

[6] Mott N. F. (1968). *J. Non-Cryst. Solids.* **1**, p. 1.

[7] Fritzsche, H. and Guevas, M. (1960). *Phys. Rev.* **119**, p. 1238.

[8] Lark-Horovitz, K. (1951). In: J. Meese (ed.), *Semiconducting Materials; Neutron Transmutation Doping in Semiconductors.* (Plenum Press, N.Y.).

[9] Shlimak, I. S. (1999). *Phys. Solid State.* **41**, p. 716.

[10] Shlimak, I. S. and Nikulin, E. I. (1972). *JETP Lett.* **15**, p.20.

[11] Efros, A .L., Lien, N. V., and Shklovskii, B. I. (1979). *J. Phys.* C **12**, p. 1023.

[12] Zhang, Y., Dai, P., Levy, M., and Sarachik, M. P. (1990). *Phys. Rev. Lett.* **64**, p. 2687.

[13] Shlimak, I., Kaveh, M., Yosefin, M., Lea, M., and Fozooni, P. (1992). *Phys. Rev. Lett.* **68**, p. 3076.

[14] Zabrodskii, A. G., Ionov, A. N., Korchazhkina, R. L., and Shlimak, I. S. (1974). *Sov. Phys. Semicond.* **7**, p. 1277.

[15] Zabrodskii, A. G., Ionov, A. N., and Shlimak, I. S. (1974). *Sov. Phys. Semicond.* **8**, p. 322.

[16] Mott, N. F. (1970). *Phil. Mag.* **22**, p. 7.

[17] Shklovskii, B. I. (1973). *Sov. Phys. Semicond.* **6**, p. 1964.

[18] Hill, R. M. (1972). *Phil. Mag.* **24**, p. 1307.

[19] Pollak, M. and Riess, I. (1976). *J. Phys.* C **9**, p. 2339.

[20] Shklovskii, B. I. (1976). *Sov. Phys. Semicond.* **10**, p. 855.

[21] Zabrodskii, A. G. and Shlimak, I. S. (1977). *Sov. Phys. Semicond.* **11**, p. 430.

[22] Aladashvili, D. I., Adamiya, Z. A., Lavdovskii, K. G., Levin, E. I., and Shklovskii, B. I. (1990). In: H. Fritzsche and M. Pollak (eds.), *Hopping and Related Phenomena* p. 283.

[23] Nguen, V. L. and Shklovskii, B. I. (1981). *Solid State Commun.* **38**, p. 99.

Chapter 3

[1] Shlimak, I. S., Lea, M. J., Fozooni, P., Stefanyi, P., and Ionov, A. N. (1993). *Phys. Rev.* B **48**, p. 11796.

[2] Shlimak, I. S. (1993). *Semiconductors.* **27**, p. 1069.

[3] Shlimak, I. S., Ionov, A. N., and Shklovskii, B. I. (1983). *Sov. Phys. Semicond.* **17**, p. 314.

[4] Ionov, A. N., Shlimak, I. S., and Matveev, M. N. (1983). *Solid State Commun.* **47**, p.763.

[5] Shklovskii, B. I. and Efros, A. L. (1984). *Electronic properties of doped semiconductors*, (Springer-Verl., Berlin; Russian original: (1979). Nauka, M.).

[6] Schoepe, W. (1988). *Z. Phys.* B **71**, p. 455.

[7] Devis, E. A. and Compton, W. D. (1965). *Phys. Rev.* **140**, p. A2183.

[8] Khondaker, S. I., Shlimak, I. S., Nicholls, J. T., Pepper, M., and Ritchie, D. A. (1999). *Solid State Commun.* **109**, p. 751.

[9] Shlimak, I., Pepper, M. (2001). *Phil. Mag.* B **81**, p. 1093.

[10] Aleiner, I. L., Polyakov, D. G., Shklovskii, B. I. (1994) in: *Proc. 22nd Int. Conf. Phys. Semicond.*(Vancouver), (World Scientific, Singapore), p. 787.

[11] Van Keuls, F. W., Hu, X. C., Jiang, H. W., and Dahm, D. J. (1997). *Phys. Rev.* B **56**, p. 1161.

[12] Khondaker, S. I., Shlimak, I. S., Nicholls, J. T., Pepper, M., and Ritchie, D. A. (1999). *Phys. Rev.* B **59**, p. 4580.

[13] Pikus, F. G. and Efros, A. L. (1994). *Phys. Rev. Lett.* **73**, p. 3014.

[14] Summerfield, S., McInnes, J. A., and Butcher, P. N. (1987). *J. Phys.* C **20**, p. 3647.

[15] Mogilyanskii, A. A. and Raikh, M. E. (1989). *Sov. Phys. JETP.* **68**, p. 1081.

[16] Shlimak, I., Kaveh, M., Ussyshkin, R., Ginodman, V., Baranovskii, S. D., Thomas, P. Vaupel, H., and Van der Heijden, R. W. (1995). *Phys. Rev. Lett.* **75**, p. 4764.

[17] Van der Heijden, R. W., Chen, G., de Waele, A. T. A. M., Gijsman, H. M., and Tielen, F. P. B. (1991). *Solid State Commun.* **78**, p. 5.

[18] Shlimak, I. S. and Lea, M. (1994). *Int. J. Mod. Phys.* B **8**, p. 891.

[19] Levin, E. I., Nguen, N. L., Shklovskii, B. I., and Efros, A. L. (1987). *Sov. Phys. JETP.* **65**, p. 842.

[20] Baranovskii, S. D., Thomas, P., and Vaupel, H. (1992). *Philos. Mag.* B **65**, p. 685.

[21] Gelmont, B. L., Gadzhiev, A. P., Shklovskii, B. I., Shlimak, I. S., Efros, A. L. (1975). *Sov. Phys. Semicond.* **8**, p. 1549.

[22] Lark-Horovitz, K. (1951) In: J. Meese (ed.), *Semiconducting Materials; Neutron Transmutation Doping in Semiconductors* (Plenum Press, N.Y. 1979).

[23] Hempshire, M. J., Wright, G. T. (1964). *Brit. J. Appl. Phys.* **15**, p. 1331.

[24] Shlimak, I. S., Efros, A. L., and Yanchev, I. Y. (1977). *Sov. Phys. Semicond.* **11**, p. 149.

Chapter 4

[1] Aleshin, A. N., Ionov, A. N., Parfen'ev, R. V., Shlimak, I. S., Heinrich, A., Schumann, J., and Elefant, D. (1988). *Sov. Phys. Solid. State*, **30**, p. 398.

[2] Aleshin, A. N., Gribanov, A. V., Dobrodumov, A. V., Suvorov, A. V., and Shlimak, I. S. (1989). *Sov. Phys. Solid State*, **31**, p. 6.

[3] Shlimak, I. S. (1990). In: H. Fritzsche and M. Pollak (eds.), *Hopping and Related Phenomena* (World Scientific, Singapore), p. 49.

[4] Terry, I., Penney, T., von Molnar, S., Besla, P. (1992). *Phys. Rev. Lett.* **69**, p. 1800.

[5] Van der Heijden, R. W., Chen, G., de Waele, A. T. A. M., Gijsman, H. M., and Tielen, F. P. B. (1991). *Solid State Commun.* **78**, p. 5.

[6] Zhang, J., Cui, W., Juda, M., McCammon, D., Kelley, R. L., Moseley, S. H., Stahle, C. K., and Szymkowiak, A. E. (1993). *Phys. Rev.* B **48**, p. 2312.

[7] Dai, P., Zhang, Y., and Sarachik, M. P. (1992). *Phys. Rev. Lett.* **69**, p. 1804.

[8] Sarachik, M. P., He, D. R., Li, W., Levy, M., and Brooks, J. S.(1985). *Phys. Rev.* B **31**, p. 1469.

[9] Shklovskii, B. I. and Efros, A. L. (1984). *Electronic properties of doped semiconductors* (Springer-Verl., Berlin; Russian original: Nauka, M. 1979).

[10] Zabrodskii, A. G., Ionov, A. N., Shlimak, I. S. (1974). *Sov. Phys. Semicond.* **8**, p. 322.

[11] Altshuler, B. I., Aronov, A. G. (1985). In: A. L. Efros, M. Pollak (eds.) *Electron-electron interaction in disordered conductors* (North Holland).

[12] Toyozawa, Y. (1962). *J. Phys. Soc. Japan.* **17**, p. 986.

[13] Ionov, A. N., Shlimak, I. S., and Efros, A. L. (1975). *Sov. Phys. Solid State.* **17**, p. 1835.

[14] Nedeoglo, N. D., Laiho, R., Lashkul, A. V., Lahderanta, E., Shakov, M. A.

(2006). *Semic. Sci. Technol.* **21**, p. 1335

[15] Agam, O., Aleiner, I. L., and Spivak, B. *ArcXiv 1312.6017v1.*

[16] Shklovskii, B. I., Spivak, B. Z. (1991). In: M. Pollak and B. Shklovskii (eds.) *Modern Problems in Condensed Matter Science. Hopping Transport in Solids.* (Elsevier), vol. 28, p. 271.

[17] Ioffe, L. B., Spivak, B. Z. (2013). *Sov. Phys. JETP.* **144**, p. 632.

[18] Shlimak, I. Khondaker, S. I., Pepper, M., and Ritchie, D. A. (2000). *Phys. Rev. B* **61**, p. 7253.

[19] Mott, N. F. (1974). *Metal-Insulator Transitions* (London, Taylor and Francis), p.124.

[20] Kurobe, A. and Kamimura, H. (1982). *J. Phys. Soc. Japan.* **51**, p. 1904.

[21] Agrinskaya, N. V., Kozub, V. I., Rentzsch, R., Lea M. J., and Fozooni, P. (1997). *JETP.* **84**, p. 814.

[22] Khondaker, S. I., Shlimak, I., Nicholls, J. T., Pepper, M., and Ritchie, D. A. (1999). *Phys. Rev. B* **59**, p. 4580.

[23] Mason, W., Kravchenko, S. V., Bowker, G. E., and Furneaux, J. E. (1995). *Phys. Rev. B* **52**, p. 7857.

[24] Shlimak, I., Friedland, K.-J., and Baranovskii, S. D. (1999). *Solid State Commun.* **112**, p. 21.

[25] Butko, V. Yu., DiTusa, J. F., and Adams, P. W. (2000). *Phys. Rev. Lett.* **84**, p. 1543.

[26] Briggs, A., Guldner, Y., Vieren, J. P., Voos, M., Hirtz, J. P., and Razeghi, M. (1983). *Phys. Rev. B* **27**, p. 6549.

[27] Ebert, G., von Klitzing, K., Probst, C., Schubert, E., Ploog, K., and Weimann, G. (1983). *Solid State Commun.* **45**, p. 625.

[28] Van Keuls, F. W., Hu, X. L., Jiang, H. W., and Dahm, A. J. (1997). *Phys. Rev. B* **56**, p. 1161.

[29] Aleiner, I. L., Polyakov, D. G., and Shklovskii, B. I. (1994). In: *Proc. Int. Conference on Physics of Semiconductors* (Vancouver), D. J. Lockwood (ed.) (Singapore, World Scientific), p.787.

[30] Fleishman, L., Licciardello, D. C., and Snderson, P. W. (1978). *Phys. Rev. Lett.* **40**, p. 1340.

[31] Baranovskii, S. D. and Shlimak, I. (1998). *ArcXiv cond-mat./9810363.*

[32] Kozub, V. I., Baranovskii, S. D., and Shlimak, I. (2000). *Solid State Commun.* **113**, p. 587.

[33] Knotek, M. L. and Pollak, M. (1972). *J. Non-Cryst. Solids.* **8-10**, p. 505.

[34] Knotek, M. L. and Pollak, M. (1974). *Phys. Rev. B* **9**, p. 644.

[35] Berkovits, R., Shklovskii, B. I. (1999). *J. Phys. Condens. Matter.* **11**, p. 779.

[36] Doetzer, R., Friedland, K.-J., Hey, R., Kostial, H., Mieling, H., Schoepe, W. (1994). *Semicond. Sci. Technol.* **9**, p. 1332.

Chapter 5

[1] Hooge, F. N. (1976). *Physica.* B+C **83**, p. 14.

[2] Kogan, Sh. M. (1977). *Sov. Phys. Usp.* **20**, p. 763.

[3] Voss, R .F. (1978). *J. Phys.* C **11**, p. L923.

[4] McCammon, D., Galeazzi, M., Liu, D., Sanders, W. T., Smith, B., Tan, P., Boyce, K. R., Brekovsky, R., Gygax, J. D., Kelley, R., Mott, D. B., Porter, F. S., Stahle, C. K., Stahle, C. M., and Szymkoviak, A. E. (2002). *Phys. Status Solidi*, B **230**, p. 197.

[5] Moseley, S. H. , Mather, J. C., and McCammon, D. (1984). *J. Appl. Phys.* **56**, p. 1257.

[6] Alessandrello, A., Beeman, J. W., Brofferio, C., Cremonesi, O., Fiorini, E., Giuliani, A., Haller, E. E., Monfardini, A., Nucciotti, A., Pavan, M., Pessina, G., Previtali, E., and Zanotti, L. (1999). *Phys. Rev. Lett.* **82**, p. 513.

[7] Shklovskii, B. I. (1980). *Solid State Commun.* **33**, p. 273.

[8] McWorter, A. L. (1957). In:*Semiconductor Surface Physics* (Univ. Pensylvania Press, Philadelphia).

[9] D'Amico, A., Fortunato, G., Van Vliet, C. M. (1985). *Solid State Electronics.* **28**, p. 837.

[10] Shlimak, I., Kraftmacher, Y. A., Ussyshkin, R., and Zilberberg, K. (1995). *Solid State Commun.* **93**, p. 829.

[11] Kozub, V I. (1996). *Solid State Commun.* **97**, p. 843.

[12] Shklovskii, B. I. (2003). *Phys. Rev.* B **67**, p. 45201.

[13] Massey, J .G. and Lee, M. (1997). *Phys. Rev. Lett.* **79**, p. 3986.

[14] Pokrovskii, V. Ya., Savchenko, A. K., Tribe, W. R., and Linfield E. H. (2001). *Phys. Rev.* B **64**, 201318.

[15] Burin, A. L., Shklovskii, B. I., Kozub, V. I., Galperin, Y. M., and Vinokur, V. (2006). *Phys. Rev.* B **74**, 075205.

[16] McCammon, D. (2005). In: *Cryogenic Particle Detection* (Springer, Berlin), p. 35.

[17] Esterman, V. (1950). *Phys. Rev.* **78**, p. 83.

[18] Low, F. J. (1961). *J. Opt. Soc. Am.* **51**, p. 1300.

[19] Zhang, J., Cui, W., Juda, M., McCammon, D., Kelley, R. L., Moseley, S. H., Stahle, C. K., and Szymkoviak, A. E. (1993). *Phys. Rev.* B **48**, p. 2312.

[20] Shlimak, I., Ionov, A. N., Rentzsch, R., and Lazebnik, J. M. (1996). *Semicond. Sci. Technol.* **11**, p. 1826.

[21] Shlimak, I. S. (1999). *Phys. Solid State.* **41**, p. 716.

[22] Kozhukh, M. L., Shlimak, I. S., Fedorov, V. V., and Yurova, E. S. (1985). *Sov. Tech. Phys. Lett.* **11**, p. 50.

[23] Vorobkalo, F. M., Zabrodskii, A. G., Zarubin, L. I., Nemish, I. Y., Shlimak, I. S. (1979). *Sov. Phys. Semicond.* **13**, p. 435.

[24] Zhang, J., Cui, W., Juda, M., McCammon, D., Kelley, R. L., Moseley, S. H., Stahle, C. K., and Szymkoviak, A. E. (1998). *Phys. Rev.* B **57**, p. 4472.

[25] De Moor, P., et al. (1993). *J. Low Temp. Phys.* **93**, p. 295.

Chapter 6

[1] Dobrego, V. P., Ryvkin, S. M. (1962). *Sov. Phys. Solid State.* **4**, p. 402.

[2] Dobrego, V. P., Ryvkin, S. M. (1964). *Sov. Phys. Solid State.* **6**, p. 928.

[3] Williams, F. E. (1960). *J. Phys. Chem. Solids.* **12**, p. 265.

[4] Hopfield, J. J., Thomas, D. G., and Gershenson, M. (1963). *Phys. Rev. Lett.* **10**, p. 162.

[5] Thomas, D. G., Hopfield, J. J., and Augustyniak, W. M. (1965). *Phys. Rev.* **140** , p. A202.

[6] Dobrego, V. P., Ryvkin, S. M., and Shlimak, I. S. (1966). *Sov. Phys. Solid State.* **8**, p. 2124.

[7] Dobrego, V. P., Ryvkin, S. M., and Shlimak, I. S. (1967). *Sov. Phys. Solid State.* **9**, p. 1451.

[8] Dobrego, V. P. and Shlimak, I. S. (1967). *Sov. Phys. Semicond.* **1**, p. 1478.

[9] Dobrego, V. P. and Shlimak, I. S. (1968). *Sov. Phys. Semicond.* **2**, p. 455.

[10] Shlimak, I. S. (1972). *Sov. Phys. Semicond.* **6**, p. 1335.

[11] Koughia, K. V. and Shlimak, I. S. (1990). In: H. Fritzsche (ed.) *Transport, correlation and structural defects* (World Scientific), pp. 213–251.

[12] Macfarlane, G. G., McLean, T. P., Quarrington, J. F., and Roberts, V. (1957). *Phys. Rev.* **108**, p. 1377.

[13] Jores, R. L. and Fisher, P. (1965). *J. Phys. Chem. Solids.* **26**, p. 1125.

[14] Reuszer, J. H., and Fisher, P. (1964). *Phys. Rev.* A **135**, p. 1125.

[15] Payne, R. T. (1965). *Phys. Rev.* A **139**, p. 570.

[16] Shklovskii, B. I. and Efros, A. L. (1984). *Electronic properties of doped semiconductors* (Springer-Verl. Berlin 1984; Russian original: Nauka, M., 1979).

[17] Reiss, H., Fuller, C. S., and Morin, F. J. (1956). *Bell Syst. Tech. J.* **35**, p. 535.

[18] Miller, A. M. and Abrahams, E. (1960). *Phys. Rev.* **120**, p. 745.

[19] Rentzsch, R. and Shlimak, I. S. (1977). *Phys. Status Solidi.* A **43**, p. 231.

[20] Shlimak, I. S. and Rentzsch, R. (1976). In: *Proc. Int. Conf. Amorphous Semicond. 76* (Balatonfuered. Publishing House of the Hungarian Academy of Sciences), p. 177.

[21] Rentzsch, R. and Shlimak, I. S. (1978). *Sov. Phys. Semicond.* **12**, p. 416.

[22] Safarov, V. I. and Titkov, A. N. (1972). *Sov. Phys. Solid State.* **14**, p. 458.

[23] Hempshire, M. J., Wright, G. T. (1964). *Brit. J. Appl. Phys.* **15**, p. 1331.

[24] Gross, E. F., Titkov, A. N., Sokolov, N. S. (1972). *Sov. Phys. Solid State.* **14**, p. 2004.

[25] Maeda, K. (1965). *J. Phys. Chem. Solids.* **26**, p. 595.

[26] Redfield, D., Wittke, J. P., and Pankove, J. I. (1970). *Phys. Rev.* B **2**, p. 1830.

[27] Kolomiets, B. T., Mamontova T. N., and Babaev, A. A. (1970). *J. Non-Cryst. Solids.* **4**, p. 2891.

[28] Davis, E. A. (1966). *Solid-State Electron.* **9**, p. 605.

[29] Davis, E. A. and Compton, W. D. (1965). *Phys. Rev.* **140**, p. A2183.